A
BRIEF
HISTORY
OF
BLACK
HOLES

About the Author

Dr Becky Smethurst is an award-winning astrophysicist and science communicator at the University of Oxford, specializing in how galaxies co-evolve with their supermassive black holes. In 2022 she was awarded the Royal Astronomical Society's Research Fellowship. Her YouTube channel, *Dr Becky*, has over 550,000 subscribers who engage with her videos on weird objects in space, the history of science and monthly recaps of space news. *A Brief History of Black Holes* is her second book; her first, *Space: 10 Things You Should Know* was named one of *BBC Sky at Night Magazine*'s books of the year, and translated all around the world.

DR BECKY SMETHURST

A BRIEF HISTORY OF BLACK HOLES

AND WHY NEARLY EVERYTHING YOU KNOW ABOUT THEM IS WRONG

PAN BOOKS

First published 2022 by Macmillan

This paperback edition first published 2023 by Pan Books
an imprint of Pan Macmillan
The Smithson, 6 Briset Street, London EC1M 5NR
EU representative: Macmillan Publishers Ireland Ltd, 1st Floor,
The Liffey Trust Centre, 117–126 Sheriff Street Upper,
Dublin 1, DO1 YC43
Associated companies throughout the world
www.panmacmillan.com

ISBN 978-1-5290-8674-4

3 5 7 9 8 6 4

A CIP catalogue record for this book is available from the British Library.

Illustrations by Megan Gabrielle Smethurst, @megansmethurst_gdesign

Other image credits: image 2 on p. 36: NASA, ESA and Allison Loll/Jeff Hester (Arizona
State University). Acknowledgement: Davide De Martin (ESA/Hubble). Image 4 on p. 43: data credit:
University of California, San Diego. Image 7 on p. 69: ESO/Landessternwarte Heidelberg-
Königstuhl/F. W. Dyson, A. S. Eddington, & C. Davidson. Image 11 on p. 129: Granger/Shutterstock.
Image 16 on p. 196: Event Horizon Telescope collaboration et al.

Typeset by Palimpsest Book Production Ltd, Falkirk, Stirlingshire
Printed and bound by CPI Group (UK) Ltd, Croydon, CR0 4YY

MIX
Paper | Supporting
responsible forestry
FSC
www.fsc.org   FSC® C116313

Visit **www.panmacmillan.com** to read more about all our books
and to buy them. You will also find features, author interviews and
news of any author events, and you can sign up for e-newsletters
so that you're always first to hear about our new releases.

To you, and your curiosity that brought you here.

*Oh, and to Mum, for always bringing me
back down to Earth with a smile.*

Contents

Standing on the shoulders of giants

At this very moment, as you sit down and relax to read this book, you are moving at an incredible speed. Earth is currently spinning on its axis, moving us through the relentless march of time from one day to the next. Simultaneously, it is orbiting around the Sun, moving us through the changing of the seasons.

But that's not all. The Sun is just one star in the Milky Way, our galaxy of over 100 billion stars. The Sun is not unique and it is not at the centre. In fact, it's fairly average and unremarkable as stars go. The Solar System is contained in a minor (seeing a pattern here?) spiral arm of the Milky Way known as the Orion Arm, and the Milky Way itself is also a fairly generic spiral-shaped island of stars – not too big, not too small.

So, this means that along with the speed of the Earth spinning, and the speed of the Earth orbiting the Sun, we are also moving around the centre of the Milky Way at a speed of 450,000 miles per hour. And what do we find at that centre? A supermassive black hole.

Yes – right now, you are orbiting a black hole. A place in space with so much material squashed in, that is so dense, that not even light – travelling at the fastest speed there is

– has enough energy to win in a tug-of-war against a black hole's gravity, once it gets too close. The idea of black holes has both captivated and frustrated physicists for decades. Mathematically, we describe them as an infinitely dense, infinitesimally small point, surrounded by an unknowing sphere from which we get no light and no information. No information means no data, no data means no experiments, and no experiments means no knowledge of what lies 'inside' a black hole.

As a scientist, the aim is always to see the bigger picture. As we zoom out of our backyard of the Solar System to encompass the whole of the Milky Way, and then even further afield to see the billions of other galaxies across the entire Universe, we find that black holes are always in the gravitational driving seat. The black hole at the centre of the Milky Way, the one currently responsible for your motion through space, is about 4 million times heavier than our Sun; which is why it's dubbed a *supermassive* black hole. While that may sound big, I've seen bigger. Once again, the Milky Way's black hole is fairly average, relatively speaking. It's not that massive, that energetic, or that active either, making it nearly impossible to spot.[1]

---

1   In fact, it made the job of figuring out that the centre of the Milky Way truly was a black hole a whole lot more difficult. If it had instead been an active black hole – one that is currently growing by 'eating' more material – it would have been one of the brightest objects in the Universe. The stars in the southern hemisphere sky would barely be visible for the glare of the Milky Way's central black hole. I think that's a world I would quite like to see.

The fact that I can accept those statements as a given, practically taking them for granted every single day, is remarkable. It was only at the end of the twentieth century that we finally realised that at the centre of every galaxy there was a supermassive black hole; a reminder that while astronomy is one of the oldest practices, carried out by ancient civilisations the world over, astrophysics – actually explaining the physics behind what astronomers see – is still a relatively new science. The advancements in technology throughout the twentieth and twenty-first centuries have only just begun to scratch the surface of the mysteries of the Universe.

Recently, I got wonderfully lost in a sprawling second-hand bookshop[2] and came across a book called *Modern Astronomy* written in 1901. In the introduction, the author, Herbert Hall Turner, states:

Before 1875 (the date must not be regarded too precisely), there was a vague feeling that the methods of astronomical work had reached something like a finality: since that time there is scarcely one of them that has not been considerably altered.

Herbert was referring to the invention of the photographic plate. Scientists were no longer sketching what they saw through telescopes but recording exactly what was seen onto huge metal plates coated in a chemical that reacted to light. In addition, telescopes were getting larger, meaning they could

---

2   Barter Books in Alnwick, Northumberland, UK. I could spend hours in there. Highly recommended.

collect more light to see fainter and smaller things. On page forty-five of my copy, there's a wonderful diagram showing how telescope diameters had increased from a measly ten inches in the 1830s to a whopping forty inches by the end of the nineteenth century. At the time of writing, the largest telescope currently under construction is the Thirty Metre Telescope in Hawai'i, which has a mirror to collect light which is, you guessed it, thirty metres across – about 1,181 inches in Herbert's money, so we've come a long way since the 1890s.

What I love about Herbert Hall Turner's book (and the reason I just *had* to buy it) is that it serves as a reminder of how quickly perspectives can shift in science. There is nothing in the book that I or my colleagues doing astronomy research today would recognise as 'modern', and I can imagine that in 120 years a future astronomer reading this book would probably think the same. For example, in 1901 the size of the entire Universe was thought to stretch to only the most distance stars at the edge of the Milky Way – about 100,000 light years away. We did not know there were other islands of billions of stars, other galaxies, out there in the vastness of the expanding Universe.

On page 228 of *Modern Astronomy*, there's an image taken with a photographic plate of what's labelled the 'Andromeda nebula'. It is instantly recognisable as the Andromeda *galaxy* (or perhaps to most people as a former Apple Mac desktop background image). Andromeda is one of the nearest galactic neighbours to the Milky Way, an island in the Universe containing of over 1 trillion stars. The image looks nearly identical to one an amateur astronomer might take from their back garden today. But even with the advancement of photographic plate technology at the end of the nineteenth century, which enabled the first images of Andromeda to be recorded, there wasn't an immediate leap to understanding what it actually was. At the time, it was still dubbed a 'nebula' – a fuzzy, dusty, not-star-like thing that was thought to be somewhere in the Milky Way, the same distance away as most stars. It took until the 1920s for its true nature as an island of stars in its own right, millions of light years away from the Milky Way, to be known. This discovery fundamentally shifted our entire perspective on our position in, and the scale of, the Universe. Overnight, our world view changed as the Universe's true size was appreciated for the first time. Humans were an even tinier drop in an even larger ocean than we had ever realised before.

The fact that we've only really appreciated the true scale of the Universe for the past 100 years or so is, in my opinion, the best example of how young of a science astrophysics truly is. The pace of advancement in the twentieth century has far exceeded even the wildest dreams of Herbert Hall Turner in 1901. In 1901, the idea of a black hole had barely crossed anyone's mind. By the 1920s, black holes were merely theoretical curiosities, ones that were particularly infuriating to physicists like Albert Einstein because they broke equations and seemed unnatural. By the 1960s, black holes had been accepted, theoretically at least, thanks in part to the work of British physicists Stephen Hawking and Roger Penrose and New Zealand mathematician Roy Kerr, who solved Einstein's general relativity equations for a spinning black hole. This led, in the early 1970s, to the first tentative proposal that at the centre of the Milky Way was a black hole. Let's just put that into context for a minute. Humans managed to put someone on the Moon before we could even comprehend that all our lives have been spent inexorably orbiting around a black hole.

It was only in 2002 that observations confirmed that the only thing that could possibly be in the centre of the Milky Way was a supermassive black hole. As someone who has been doing research on black holes for less than ten years, I often need reminding of that fact. I think everyone has a tendency to forget the things that, even up until recently, we didn't know. Whether that's what life was like before smartphones, or that we have only been able to map the entire human genome this millennium. It's understanding the history of science that allows us to better appreciate the knowledge we now hold dear. A look back into science history is like riding the collective train of thought of thousands of researchers. It puts into perspective those theories that we are so used to parroting we forget the fire in which they were first forged. The evolution of an idea helps us to understand why certain ideas were discarded and some were championed.[3]

---

3   As someone who loves science history, it is both painful and curiously fascinating to watch the rise of 'flat-Earthers', who claim that the Earth is flat. They insist that NASA and the US government (with all other space agencies and governments presumably in cahoots) have been perpetuating the lie of a spherical Earth. What's interesting is that the thoughts and arguments they discuss among themselves are the very same ones that early Greek philosophers had thousands of years ago,

It's a thought I have a lot when people challenge the existence of dark matter. Dark matter is matter that we know is there because of its gravitational pull, but we cannot see it because it does not interact with light. People question how plausible it really is that we're unable to see what we think makes up 85 per cent of all the matter in the Universe. Surely there must be some other thing we've not yet thought of? Now, I would never be so arrogant as to claim that we have indeed thought of absolutely everything, because the Universe is constantly keeping us on our toes. But what people forget is that the idea of dark matter didn't just pop up fully formed one day to explain away some curiosity about the Universe. It came about after over three decades worth of observations and research pointed to no other plausible conclusion. In fact, scientists dragged their feet for years, refusing to believe that dark matter was the answer; but in the end the evidence was just overwhelming. Most observationally confirmed scientific theories are shouted about from the rooftops; dark matter, however, must have been the most begrudgingly agreed upon theory in all of human history. It forced people to admit we knew far less than we thought we did, a humbling experience for anybody.

That's what science is all about: admitting the things we

---

but eventually discarded after more experiments and observations. This is the crucial part that 'flat-Earthers' struggle with – letting go of an argument they are emotionally attached to when their experiments show them that the Earth is not flat. They somehow cannot end their odyssey of confirmation bias. A society never progresses if it refuses to change its beliefs when presented with overwhelming evidence to the contrary.

don't know. Once we do that, we can make progress, whether for science, for knowledge, or for society in its entirety. Humanity as a whole progresses thanks to advancements in knowledge and in technology, with the two driving each other. A thirst for more knowledge about the size and contents of the Universe, to see further and fainter things, drove the advancement of telescopes (from forty inches across in 1901 to thirty metres across in 2021). Tired of cumbersome photographic plates, the invention of digital light detectors was pioneered by astronomers, and now we all carry a digital camera around in our pockets. That invention saw improvements to image analysis techniques, which were needed to understand the more detailed digital observations. Those techniques then fed into medical imaging, such as MRIs and CT scanners, now used to diagnose a whole host of ailments. Getting a scan of the inside of your body would have been unimaginable a mere century ago.

So, like all scientists, my research on the effects of black holes stands on the shoulders of the giants who have come before me: the likes of Albert Einstein, Stephen Hawking, Sir Roger Penrose, Subrahmanyan Chandrasekhar, Dame Jocelyn Bell Burnell, Sir Martin Rees, Roy Kerr and Andrea Ghez to name but a few. I can build upon the answers that they worked so hard and so long for, to pose new questions of my own.

It has taken over 500 years of scientific endeavours to just scratch the surface of what black holes are. It's only by delving into that history that we can hope to understand this strange and enigmatic phenomena of our Universe, one we still know so little about. From the discovery of the smallest, to the largest; the possibility of the first black hole, to the last; and why they're even called black holes in the first place. Our jaunt through science history will take us on a journey from the centre of the Milky Way to the edges of the visible Universe, and even consider the question that has intrigued people for decades: what would we see if we 'fell' into a black hole?

To me, it's incredible that science can even hope to answer questions like that, while simultaneously surprising us with something new. Because, while black holes have long been thought to be the dark hearts of galaxies, it turns out they're not 'black' at all. Over the years, science has taught us that black holes are in fact the brightest objects in the entire Universe.

I

Why the stars shine

The next time you have a clear night, with no clouds spoiling the view, stand with your eyes closed for a few minutes by the door to outside. Before you step out and look up, give your eyes time to adjust to the darkness. Even young children notice how when you first turn the bedside light off before sleep, the room plunges into pitch blackness. But wake in the middle of the night and you can see shapes and features again in even the lowest of lights.

So if you want to truly be awed by the night sky, let your eyes take a break from the bright lights of home first. Let your night vision develop and you won't be disappointed. Only once your eyes are primed and ready can you then step outside and change your perspective on the world. Instead of looking down, or out, look *up* and watch thousands of stars burst into view. The longer you stand in the darkness, the better your night vision will be and the more stars will pepper the sky with tiny pinpricks of light.

As you gaze skyward, you might spot things you recognise, such as shapes in the patterns of stars that we call constellations, like Orion or the Plough.[4] Then there'll be things that

---

4   Also known as the Big Dipper.

aren't familiar. But by just gazing at the sky and noting the brightness or perhaps the position of a star, you join an incredibly long list of humans from civilisations the world over, both ancient and new, that have done the very same and found themselves awed by the beauty of the sky. The stars and planets have long held an important cultural, religious or practical role in society. From navigation by land or sea, to helping people keep track of the seasons, leading to the development of the first calendars.

In the modern world, we have lost that innate connection with the night sky, with many of us not able to notice how the stars change with the seasons or pick out visiting comets because of the ever-present light pollution in cities drowning them all out. If you're lucky enough to live somewhere you can see the stars, perhaps you might notice how the position of the Moon changes from night to night, or that one particularly bright 'star' wanders across the sky as the months go by. The Greeks also noticed these 'wandering stars' and dubbed them just that: planētai, meaning wanderer (the root of the modern English word, planet).

But not all of us can just look up and enjoy the view for what it is. Some of us want answers; an explanation of the things we see in the sky. It's natural human curiosity. The very nature of what stars are and how they shine were questions that plagued humanity for centuries. In 1584, Italian philosopher Giordano Bruno was the first to suggest that the stars themselves might be distant Suns, even going so far as to suggest that they may also have planets of their own orbiting them. This was an idea that was incredibly controversial at the time, and came just forty-one years after the neat mathematical idea of the Sun, and not the Earth, being the centre of the Solar System was published by Polish mathematician and philosopher Nicholas Copernicus. Copernicus was a big fan of the simplicity and mathematical beauty of circles, and thought that if you arranged the Solar System with the Sun at the centre and the planets moving around it on circular paths, that would be the most mathematically beautiful way of arranging things. He wasn't serious about it astronomically, necessarily, he just enjoyed the geometry of the whole idea.

But after a few more decades, there were those that started to support the idea astronomically, like Bruno and his fellow Italian astronomer Galileo Galilei, who would both eventually be punished for this supposed heresy against Catholic doctrine. It would take the combined efforts of Tycho Brahe, Johannes Kepler and Isaac Newton over the next century or so to compile overwhelming evidence in favour of the Sun being at the centre of the Solar System, and for the idea to finally be accepted both scientifically and publicly following the publication of Newton's *Principia* in 1687. First, Newton determined the laws of gravity and the movements of the planets in their orbits. The same force that keeps us trapped here on the Earth's surface is what causes the Moon to orbit the Earth and the Earth to orbit the Sun. These roughly circular orbits of planets around the Sun explained why the planets appeared to move backwards night after night in the sky for parts of the year, a phenomenon known as retrograde motion. Those planets closer to the Sun appear to be moving backwards in the sky when they were on the other side of the Sun (like cars on the opposite side of a circular

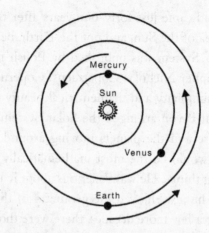

Mercury in 'retrograde' appears to be moving backwards,
but it's just on the other side of the 'racetrack'.

racetrack),[5] and those planets further out would appear to move
backwards as Earth overtook them as it moved faster in its orbit.

While Bruno was ahead of his time, his idea that the Sun
was a star like any other, albeit a lot closer, still didn't help
to reveal how they shine. However, realising that the Sun was
at the centre of the Solar System and governed by the same
forces that we experience here on Earth removed the Sun's
God-like status, rendering it something more ordinary in
people's minds. Physicists of the 1700s started wondering
whether the Sun and the stars could be powered by everyday
processes like combustion, going as far as considering whether

---

5   So stop blaming your problems on 'Mercury in retrograde'. Mercury
is just happily orbiting the Sun like it has done for the past 4.5 billion
years. Earth's perspective on the position of inanimate rocky objects
doesn't have any influence on your life.

burning coal could account for the amount of energy outputted as light. Spoiler alert: it can't. If the entire Sun was made of coal, it would burn through it at its present rate of energy production in just 5,000 years.[6] Given that recorded history went back further than that – the Great Pyramid of Giza had been built over 4,000 years earlier – and that the Earth was then thought to be 6,000 years old, this idea was eventually dismissed.

So, if the Sun wasn't made of coal, then what was it made of? Figuring out what the Sun was made of became a huge focus of physicists in the 1800s, but it was a Bavarian glassmaker who made the first breakthrough. Joseph Ritter von Fraunhofer was born in 1787, the youngest of eleven children in a family boasting many generations of glassmakers. His story has all the tropes of a good Disney movie: by the time he was a teenager, he was an orphan sent to apprentice with a master glassmaker in Munich who made decorative mirrors and glass for the royal court. His master was cruel to him though, depriving him of an education and a reading lamp to read his precious science books after dark. But one night his master's house collapsed, burying Joseph alive inside. This was such huge news in the city of Munich that a Prince of Bavaria even came to the scene of the disaster and was there as Fraunhofer was pulled alive from the rubble. When the prince heard of Joseph's plight, he set him up with a new master in the royal palace who supplied

---

6 Desperate to resolve this problem, scientists of the time even considered the idea that meteors impacting with the Sun could bring extra coal deposits to keep it going for longer.

him with all the books on mathematics and optics that he could get his hands on. A true fairy tale story.

But the story doesn't finish there, Fraunhofer ended up working at the Optical Institute in Benediktbeurern, where he was put in charge of all glassmaking, improving methods for grinding super-smooth glass for use as lenses in tele-scopes. The problem that Fraunhofer applied himself to was understanding the pesky refractions (a change in direction of the light) that would occur through the glass, scattering some of the light into the colours of the rainbow. This made his lenses imperfect. He was trying to measure how much the light was refracted, i.e. how much its direction was changed, through different types and shapes of glass. Isaac Newton had already shown in the 1600s that white light was made up of all the colours of the rainbow, showing how refraction occurs through a prism, changing the direc-tion of red light less, and changing the direction of blue light more, to reveal the rainbow. If you're picturing Pink Floyd's *Dark Side of the Moon* album cover you're on the right track.

The problem Fraunhofer had was that the colours of the rainbow aren't clearly separated from each other. Next time you see a rainbow in the sky, see if you can pick out where the green ends and the blue begins: it's impossible to tell. The colours blend into each other, making something really pleasing to the eye, but incredibly frustrating if you're trying to measure how much the direction of each colour of light is changed by. So, Fraunhofer started experimenting with different sources of light. He noticed that when he used the light of a flame burning sulphur there was one section of the rainbow, of a yellow-orange colour, that was much brighter than the rest. He became curious whether the Sun also showed this bright yellow patch in the light as well, tweaking his experiment to change the path of the light more and more to get the rainbow to cover a larger area: he essentially managed to 'zoom in' on the rainbow to see more detail. By doing so, he invented the very first spectrograph; an instrument that is the cornerstone of modern astronomy and astrophysics.

Fraunhofer was shocked at what he then saw using his spectrograph with light from the Sun; instead of brighter patches of light, he noticed there were some colours of light from the Sun that were missing entirely. Dark lines in the rainbow, gaps that no one else had spotted before. He labelled the ten most obvious dark sections at first, eventually recording 574 gaps in the rainbow of light from the Sun. If you could zoom in on a rainbow in the sky, this is always what you would see.

Intrigued by this finding, Fraunhofer investigated further, finding that the gaps appeared in sunlight reflected off the Moon and the planets, and objects on Earth. He wasn't certain though whether the gaps in the light were a true property of sunlight, or caused when the light passed through the Earth's atmosphere. So he then used his spectrograph to look at the light from other stars, like the bright star Sirius, the 'dog star'[7] near the constellation of Orion (Orion is supposed to look

---

7 Yes, that is where Sirius Black got his name.

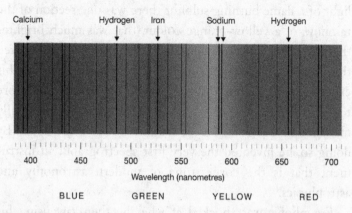

Calcium Hydrogen Iron Sodium Hydrogen

400   450   500   550   600   650   700

Wavelength (nanometres)

BLUE   GREEN   YELLOW   RED

The Sun's rainbow split by a spectrograph showing the missing colours that Fraunhofer spotted. Eventually Bunsen and Kirchoff showed that they were caused by elements in the Sun absorbing these colours, revealing what the Sun was made of.

like a hunter, with a smaller constellation next to him of his hunting dog, of which Sirius is the brightest star). Fraunhofer noticed that the gaps appeared once again in the light from Sirius, but they were in completely different locations, with a different pattern to sunlight. He concluded that it wasn't the Earth's atmosphere causing these gaps, but something in the very nature of stars themselves.

With this discovery in 1814, Fraunhofer essentially kick-started modern astrophysics as we know it, and he lived happily ever after. Or at least, that's how the Disney movie of Fraunhofer's life would end. In reality, he died from tuberculosis at only thirty-nine years of age in 1826. The glass furnaces he worked with contained poisonous lead oxide and most likely contributed to his death.

Fraunhofer's untimely death meant that he never lived to see these gaps in the Sun's rainbow of light explained a few decades later, in 1859, by German physicist Gustav Kirchoff and chemist Robert Bunsen. Kirchoff and Bunsen didn't set out to explain what Fraunhofer had seen, but were instead investigating something else using Bunsen's new invention, which produced a very hot, sootless flame (that also wasn't blindingly bright) for use in the laboratory. Today, every science lab around the world has one, from high-tech research institutes to school chemistry classrooms: a Bunsen burner.

Using a Bunsen burner, Kirchoff and Bunsen would burn various different elements in the flame and record what colour of light was given off. They even used a newly updated version of Fraunhofer's spectrograph to split the light given off into its component colours. They found that each element burned with a very specific colour, or wavelength of light. For example, sodium burns a bright yellow colour, with a wavelength of exactly 589 nanometres (0.000000589 m), which is the colour of old-fashioned yellow street lamps which use sodium-powered bulbs. Kirchoff noticed that one of the missing gaps in the Sun's rainbow of light that Fraunhofer had recorded was also at exactly 589 nanometres. Could it be that sodium was also present in the Sun, but instead of emitting light of that colour, it was absorbing it?

Kirchoff and Bunsen then cross-referenced all the wavelengths emitted by elements they had categorised in their lab with those recorded by Fraunhofer and found matches everywhere, suggesting the Sun contained sodium, oxygen, carbon, magnesium, calcium, hydrogen, and many other elements. This essentially confirmed that the Sun was indeed made of the same elements that we find on Earth. In his honour, Kirchoff and Bunsen dubbed the gaps in the Sun's rainbow of light 'Fraunhofer lines'.

So, in 1859, the problem of what the Sun was made of was solved, but the problem of how the Sun was powering itself with the same elements that made up Earth was still unsolved. There's a wonderful *Scientific American* article from August 1863 entitled: 'Experts Doubt the Sun is Actually Burning Coal', which states:

> *The sun, in all probability, is not a burning, but an incandescent, body. Its light is rather that of a glowing molten metal than that of a burning furnace.*

In other words, it's something like Earth, but for some reason much hotter, so much so that it is glowing.

This article was based on the work of British physicist William Thompson – who would later be dubbed Lord Kelvin as he became the first scientist to be elevated to the House of Lords (the scientific unit of temperature, the kelvin, is named in his honour) – and German physicist Hermann von Helmholtz. Kelvin and Helmholtz are giants in the world of thermodynamics: they pioneered our understanding of heat and temperature. In 1856, Helmholtz published his idea that the Sun generated heat because it was being squished under gravity, essentially transferring huge amounts of energy from the crush of gravity inwards, into kinetic energy, which gives atoms (the building blocks of all the elements) more energy to move faster, heating up the Sun so that it glows like a hot piece of metal or molten glass.

In 1863, Kelvin used Helmholtz's idea to calculate that the Sun would be able to power itself this way for at least 20 million years – far longer than the Earth's supposed 6,000-year age that had stumped the 'Sun is powered by coal' calculation. The same year, Kelvin also applied the ideas of heat transfer to the Earth to calculate its age, by assuming that the Earth was once molten and has since been cooling for long enough to give us a solid crust of rock to stand on. Kelvin calculated that the Earth must also be around 20 million years old.[8] The similarity of Kelvin's two estimates, the age of the Sun and the age of the Earth, were interpreted as a success. If the Earth and the Sun formed at the same time, out of the same mix of elements, then this would finally explain the similarity of elements shared between the Earth and the Sun, and solve the problem of what powered the Sun, in one fell swoop.

The physicists, therefore, were happy, but the biologists and geologists were most definitely not. Because a few short years before Kelvin made his age estimates, in 1859, a biologist called Charles Darwin had published his book *On the Origin of Species*, detailing his new theory of evolution. In it,

---

8   This is *way* off the current estimate of the age of the Earth – around 4.5 billion years. Kelvin didn't know to take into account the heat given off by radioactive decay in the Earth's core, as radioactivity hadn't been discovered yet.

he said that all life on Earth had evolved from a common ancestor, branching through different mutations motivated by natural selection (what Herbert Spencer would call 'survival of the fittest' a few years later). By the 1870s, the majority of the scientific world – and members of the public who were paying attention – had accepted the idea of evolution. There was just one problem: this process of evolution took time, a lot of time. Darwin himself in his 1872 edition of *On the Origin of Species* commented that Kelvin's 20 million-year estimate for the age of the Earth wouldn't give enough time for evolution to occur. Evolution needs billions, not millions, of years.

Meanwhile, the geologists were attempting to use their own methods to calculate the age of the Earth. Either by working out the rate that rocks form and lay down sediment, or by considering the build-up of salt in the oceans. The chap who had this idea was Irish geologist and physicist John Joly; in 1899 he reasoned that salt (i.e. sodium chloride) dissolves out of rocks, into rivers, which then meet the sea. If the Earth's oceans originally formed with no salt in them, then from the rate at which salt flows through rivers, you can work out how long it would have taken for salt to build up to the concentrations we measure in the sea today, and therefore get an estimate for the age of the Earth. In case you're wondering, Joly estimated that there are 14,151 trillion tons of sodium in the ocean, whereas in rivers there are 24,106 tons of sodium per cubic mile of water. He also estimated that the total volume of water that leaves rivers and enters the ocean is 6,524 cubic miles per year. Running through the maths gives you an estimate that the build-up of salt in the ocean took almost 90 million years.[9]

This was closer to what the biologists were expecting – still not the billions of years that would be the boon for Darwin's theory of evolution, but the death knell for Kelvin's estimate of the age of the Sun. Another breakthrough was kick-started in 1895, when French physicist Henri Becquerel discovered that uranium atoms were unstable, and would spontaneously transform into more stable elements over time, giving off radiation in the process. His PhD student, French-Polish physicist and chemist Marie Skłodowska-Curie, decided to investigate this radiation for her PhD thesis, using a tool that her husband, Pierre Curie (studying crystals at the time), had invented fifteen years earlier to measure electric charge. She found that the radiation given off by the uranium atoms caused the air around them to conduct electricity, and hypothesised that the radiation must come from the atoms themselves (rather than be caused by an interaction with air molecules).

After the birth of her daughter Irène in 1897, Curie dedicated herself to finding yet more unstable elements, discovering thorium and finding it produced four times more radiation than uranium. By 1898, her husband Pierre had abandoned his own work on crystals for Marie's far more

---

9   $14,151,000,000,000,000 / (24,106 \times 6,524) = 89,980,422$ years. It's worth noting that this gives the wrong answer because of *many* incorrect assumptions. For one, the rate at which salt flows through rivers is not constant with time, and for another, because the oceans have long been in a steady state of salinity: rocks on the ocean floor absorb salts as quickly as they are pumped in by rivers.

interesting research on this unknown radiation. By the end of the year they had announced the discovery of two more unstable elements, which they dubbed polonium in honour of Marie's homeland of Poland, and radium, after the Latin word for 'ray'. In doing so they coined the phrase 'radioactivity'. In 1903, Marie and Pierre Curie and Henri Becquerel were awarded the Nobel Prize in Physics for their discovery and characterisation of radioactivity.[10]

What's so key about the discovery of radioactivity is that it establishes that the transformation (or 'decay') of unstable elements happens at a constant rate. If you can measure the amount of the unstable element and compare it to the stable element it decays into, then you can work out how long it has been decaying for. This was the breakthrough that revolutionised geology. By 1907, this method of 'radioactive dating' had been applied to the rocks of Earth, suggesting Earth (and therefore the Sun it orbited around) was at least a few billion years old.[11]

Finally, a value that made sense for all the biologists long convinced by Darwin's theory of evolution. But it was a value that caused more pain for physicists trying to determine how

10 Initially, only Pierre Curie and Henri Becquerel were to be awarded the prize, but a committee member, Swedish mathematician Magnus Gösta Mittag-Leffler, alerted Pierre to the situation, who promptly complained, and Marie Curie's name was rightly added to the prize. A lesson for us all in how to be an ally.

11 Modern radioactive dating measurements estimate that the Earth is 4.55 billion years old (with an uncertainty of around 0.05 billion years, or 1 per cent).

the Sun could possibly be shining, finally scrapping Kelvin's ideas. Although radioactivity produces heat (and is enough to explain the heat given off by Earth), it isn't nearly enough to be the sole source of energy in the Sun. So, at the start of the twentieth century, we had a good idea for how old the Sun was (at least as old as the Earth), but had no idea how it could possibly have been shining for that long.

Enter stage left: German physicist Albert Einstein. Along with Stephen Hawking, Einstein's name is perhaps most synonymous with black holes. He is perhaps the grandfather of black holes themselves, with his theories kick-starting decades of research into the nature of gravity, space and time. But for this part of the story, we need only his most famous equation (arguably the most famous equation *ever*), which he proposed in 1905: $E = mc^2$. E stands for energy, m for mass and c for the speed of light – a whopping 299,792,458 metres per second. It means that energy and mass are *equivalent* – they are essentially the same thing and intrinsically linked. Mass can be converted into energy.[12] Here, finally, was something that could explain where the huge amounts of energy produced in the Sun for billions of years was coming from; it was converting its enormous mass directly into energy. But how?

The first clue came in 1919 from French physicist Jean Baptiste Perrin, who would go on to win the 1926 Nobel Prize in Physics for showing that individual atoms could join together to make molecules. For example, O_2 is formed of

---

12 Which also explains why radiation is produced when a heavier unstable element radioactively decays into a lighter stable one.

two oxygen atoms joined together. In his work studying atoms and molecules, he discovered that a helium atom, with four particles, weighs *less* than the total mass of four hydrogen nuclei, with one particle each. The mass difference was tiny, at just 0.07 per cent, but with $E = mc^2$, a tiny mass can turn into a huge amount of energy. Perrin[13] realised the significance of what he had found and suggested that this could be what is powering the Sun. If four atoms of hydrogen could be brought together to make helium, the leftover mass could become energy given off as light. The problem was that Perrin didn't have a physical model for how this actually happened, pointing out that the central nuclei of hydrogen atoms were positively charged and would repel each other with a huge force (atoms have a central nucleus with positively charged particles, orbited by smaller negatively charged particles known as electrons).

It would take the stubbornness of English physicist Arthur Eddington in 1920 to convince the world that if this process of *fusing* four hydrogen nuclei together to make helium was going to happen anywhere, then it had to be happening in stars. Eddington was already somewhat of a household name by 1920, after writing a number of articles explaining Einstein's newest theory of general relativity to the English-speaking world (more on that later). His own research, though, was on the nature of stars, and in 1920 Eddington reasoned a few

---

13 To my fellow *Wheel of Time* fans; no, I can't read this paragraph without giggling either. Perrin Aybara: blacksmith, wolfbrother and nuclear physicist.

things: first, using the same methods as Lord Kelvin himself, that the temperature at the centre of stars would be around 10 million degrees Celsius, and that at these temperatures our understanding of the interaction of nuclei and repulsive forces keeping positively charged hydrogen nuclei apart might break down. Second, that only 5 per cent of the mass of the Sun had to be hydrogen to produce enough energy to keep it burning for the billions of years that Earth had been around. These were all ideas that were proved correct over the next few decades, and further contributed to Eddington's status as a BNIP (a Big Name in Physics).

In 1925, British-born American astronomer Cecilia Payne-Gaposchkin published her PhD thesis. Her research showed how Fraunhofer's gaps in the rainbow of light from the Sun meant that hydrogen outweighed every other element in the Sun by a million times. Far more than just 5 per cent of the Sun was made of hydrogen. The final piece of the puzzle came in 1928, when American-Russian physicist George Gamow ran through the maths and realised there was a vanishingly small probability of a hydrogen nucleus outwitting the electric repulsion between it and another hydrogen nucleus to allow them to fuse together. The probability might be incredibly small, but crucially, *it isn't zero*. So, if you have enough hydrogen squished into one place, like in the Sun, then theoretically this skipping of the repulsion can occur enough times to produce enough energy so that the Sun shines.

Finally, the problem was solved. Hydrogen was the fuel of the Sun and all the stars in the night sky: nuclear fusion was what made them shine. I can't help but wonder how much of that story we'd even know if we couldn't see the stars. Would we have even thought to ask questions like 'what makes the stars shine?' Would we have realised what the Sun actually was? Perhaps if Earth was in orbit around two stars, so that it was daytime on both sides of the planet, we would have had endless daytime and never have seen the night sky. What questions would we have never known to ask? What advancements in knowledge and technology would have eluded us?

I think we as humans have a lot to thank for the curiosity that staring up at the night sky has to offer. Not least for our knowledge of my favourite thing: black holes. Because once we figured out how the stars shine, this inevitably led us to another question: what happens when the fuel runs out? What happens when a star dies? And it is this simple question that eventually leads us to a black hole.

2

Live fast, die young

In AD 1054, a star in the constellation of Taurus (named by the Greeks for its apparent likeness to a bull[14]) flared dramatically in brightness; so much that it could even be seen during the day, when the Sun is bright enough to outshine all other stars. Chinese astronomers referred to these brightened stars as 'kèxīng' (客星) – 'guest stars' – and meticulously recorded their appearance. They noted that the guest star of 1054 was visible for another 642 nights in the night sky (around twenty-one months!), before fading away entirely.

Today, almost a thousand years later, if you were to take a telescope and look at that very same position in the sky in the constellation of Taurus, you would see something dramatically different from a star: you would see a nebula. A maelstrom of gas and dust lit from the centre by the glowing embers of a star too faint to see. This is the leftovers of a dead star, one that ran out of hydrogen fuel and as it desperately tried to prevent the inevitable, outshone every other star in the sky for those short few months, before leaving behind a shadow of what it once was. This ghostly scene is known

14 Although these are also the Greeks who thought that a 'W' shape resembled a woman sat on a throne, so named it after Queen Cassiopeia.

The Crab Nebula; the remnant of supernova SN 1054.

as the Crab Nebula, and it is a milestone in the history of humanity's knowledge of the death of stars, and our realisation of the existence of black holes.

While the Crab Nebula is not one of the brightest objects in visible light that we can see with our eyes, it is one of the brightest emitters of incredibly high-energy light known as gamma rays. Light comes in all different shapes and flavours, which are determined by the amount of energy the light wave has. In visible light (also known as optical light), the different colours that we can see with our eyes are due

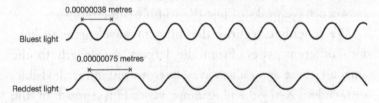

0.00000038 metres

Bluest light

0.00000075 metres

Reddest light

The different wavelengths of red and blue light.

to different wavelengths of light. Blue colours are more energetic – there are more waves that arrive each second – whereas red colours are less energetic, with fewer waves arriving each second. The number of waves that arrive each second is measured as the frequency, or alternatively you can measure it as the distance between the peaks of the waves, known as the wavelength.

Our eyes can only detect colours that have waves separated by just 0.00000038 metres for blue light and 0.00000075 metres for red light (a frequency range of 790–400 trillion waves per second). 'White' light, like that from a torch or the Sun itself, is a mixture of all the colours; as demonstrated beautifully by a rainbow. As light from the Sun passes through drops of water in the air, it is split into its component colours for us all to marvel at. What's wonderful to think about when you see a rainbow is that you're barely even seeing the full picture. There are colours beyond the red at the top of the rainbow and the blue at the bottom; but they're colours that our eyes can't see. The Sun doesn't just emit visible light, it emits light of all wavelengths, from the least energetic light, the laziest, with huge, kilometres-long separations between wave peaks, to the most energetic light, with tiny separations between wave peaks of just the width of an atom.

We roughly categorise the different wavelengths of light into different types. From the largest wavelength to the smallest, these are: radio waves; microwaves; infrared; visible; ultraviolet; X-rays; and gamma rays. This spread of the different wavelengths of light is the true full spectrum of light; the entire rainbow, of which we only get the tiniest glimpse. Despite not being able to see these wavelengths of light, this hasn't stopped us from exploiting them for our gain. From using radio waves to communicate, microwaves to cook our food, infrared light in our TV remotes, ultraviolet for killing bacteria, X-rays for getting a glimpse inside our bodies, and gamma rays in radiotherapy to fight cancer.

However, the more energetic the light, the more dangerous it becomes to life here on Earth. Thankfully, the Earth's atmosphere filters out the majority of the wavelengths of light produced by the Sun. The highest-energy ultraviolet light is absorbed by oxygen atoms in our atmosphere, producing the ozone layer. Similarly, oxygen and nitrogen atoms absorb all the X-rays and gamma rays, and moisture in the atmosphere absorbs microwaves. The only light that makes it to the ground is visible light, some ultraviolet (which can burn our skin – sunburn) and harmless radio waves. The Sun is 10 million times brighter in visible light compared to radio waves, so it's no surprise that human eyes evolved to see the bright kind of sunlight which actually makes it to the ground. Perhaps on another planet, with a different type of atmosphere, our eyes would be able to detect a completely different part of the spectrum of light, with brand new colours that we don't even have a hope of visualising.

As astronomers, though, we are no longer limited by the puny sensitivity of the human eye. We have 'evolved' a step further, developing detectors sensitive to different types of light. The problem is Earth's pesky atmosphere which, while protecting life from harmful radiation, also scuppers any detection of X-rays coming from the vastness of space. So we strap our X-ray detectors to telescopes and launch them up into orbit around Earth, beyond the atmosphere that blocks our view. With these telescopes we've been able to open our eyes to see the tiny pinpricks of light dotting the infrared, X-ray, and gamma-ray sky that for so long remained hidden to us. Including light from the Crab Nebula, which might have outshone the Sun in visible light back in 1054, but now outshines the Sun and nearly everything else in the sky in gamma rays.

It's the different colours and types of light we receive from stars that let us work out how hot they are, what type they are and what will happen to them when they die. There are some stars, like Betelgeuse in the constellation of Orion, that appear slightly reddish in colour; you can even see this with the naked eye when the sky is dark, but it's even more obvious if you snap a picture (even a ten-second 'night mode' shot on most smartphones will reveal this if you can't quite make it out with your eye). Similarly, there are some stars, like Sirius, which appear blueish in colour.

So, using the light from stars, astronomers decided to do what all good scientists do and classify them. With a system. In the same way biologists have their classification of the animal kingdom and chemists have their periodic table, astronomers have their star classification system. This was made possible by Fraunhofer's invention of the spectrograph – splitting the light from a star into its rainbow reveals the gaps where light is missing; the hidden fingerprints revealing what that star was made of. Because, as Fraunhofer had pointed out, not all stars had the same pattern of missing colours as the Sun.

It was this observation that allowed Italian astronomer Angelo Sechhi to sort stars for the first time into three broad categories. In 1863, Secchi began recording the spectrum of light from different stars, just like Fraunhofer first did with the Sun, amassing over 4,000 of them to analyse. He realised that although the patterns of the missing colours varied slightly from star to star, they could be roughly sorted into three categories, which he called I, II and III using Roman numerals (he also ended up adding two more rarer classes to his scheme, IV in 1868 and V in 1877). The Sun is a type-II star according to Secchi, which means it has a lot of missing colours. We now know these missing colours correspond to the Sun having lots of heavier elements like carbon, magnesium, calcium and iron absorbing light at those colours – we call these metal lines: any element heavier than hydrogen is classed as a 'metal' by astronomers, much to all chemist's chagrin.

Secchi wasn't the only one interested in classifying stars based on their light. In the 1880s, American astronomer and director of the Harvard College Observatory, Edward Pickering, also turned his attention to classifying stars. Pickering amassed over 10,000 star spectra to analyse, but he didn't do it alone. Pickering had help from the 'Harvard computers'. The word computer today refers to a machine, but in Pickering's day computers were people: 'one who computes'. Teams of people were hired to do repetitive, tedious jobs and incredibly complex mathematical calculations. These computers were more often than not women, who would make discoveries in the data they had been given to process or glean insight that had previously been missed.[15] At the Harvard College Observatory men would do the manual labour of moving the telescope and taking the images or spectra on large photographic plates, then the women would do the tedious, repetitive cataloguing of the brightness or spectrum of a star. By today's definitions, the men were doing the astronomy, the women doing the astrophysics.

Harvard computer Williamina Fleming did the bulk of the classification of Pickering's 10,000-star spectrum (discovering ten new 'guest stars' in the process), and together Pickering and Fleming revised Sechhi's system to have more specific classes. They split his five broad classes (I–V) into sub-classes using the letters A–Q, giving seventeen different types of stars in total. As you went through the alphabet the amount of absorption by hydrogen decreased. The work was published in 1890 and was known as the *Draper Catalogue of Stellar Spectra*, having been funded by Mary Anna Palmer Draper, the widow of American doctor and keen amateur astronomer Henry Draper.

---

15 I can also highly recommend the 2017 film *Hidden Figures*, which celebrates the contributions of black women computers at NASA during the space race and the Apollo missions, in particular Katherine Johnson, Dorothy Vaughan and Mary Jackson.

This method of classifying stars was deemed overly complicated by some, especially another Harvard computer called Annie Jump Cannon. In 1890, Harvard College Observatory branched out from studying stars only in the northern hemisphere sky and built an observatory in Arequipa, Peru to get data from (the many more) stars in the southern hemisphere sky. Cannon was tasked with classifying all the stars in the southern sky down to a certain brightness for a revised version of the Draper catalogue. While doing so, she also simplified the classification system, sticking to the alphabetised letters but dropping all but A, B, F, G, K, M and O. She noticed that most stars were a mix of two types, a halfway house between the A and B type, for example. So, instead of seventeen individual types, she added a number between 0 and 9 to specify if the star was between two types, e.g. an A5 star. The Sun is a G2 star in Cannon's system, Sirius with its blueish colour is an A1 and Betelgeuse with its reddish colour is an M2 star.

Pickering and Cannon first published this system in 1901, but the work didn't end there. The Draper catalogue was not yet complete, with many more stars in the sky yet to be classified. The full catalogue of 225,300 stars was published in volumes between 1918 and 1924; in working towards it Cannon and her computer colleagues at the observatory were classifying the spectra of over 5,000 stars *per month* using her system.

So by the early twentieth century, astronomers had a system for *sorting* stars, but understanding *why* they could be sorted that way would take that little bit longer. What made the spectrum of stars look different? What made them shine with slightly different colours? Well, during the completion of the Draper catalogue in 1911, Danish chemist and astronomer Ejnar Hertzsprung had worked out the distance to some of the stars listed. With the distance he could work out their actual brightness, rather than the brightness they appear to be from Earth, and noted that the actual brightness was proportional to the amount of light stolen away in the absorption lines (the absorbed wavelength/colour doesn't disappear completely, but is very faint compared to the overall amount

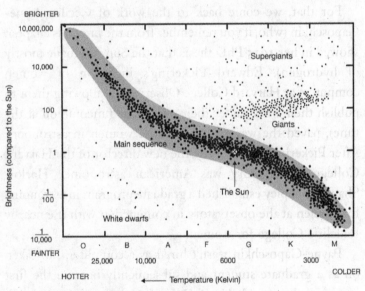

The Hertzsprung–Russell diagram for nearby stars. The 'main sequence' of stars is the correlation originally seen by Hertzsprung and Russell and is where normal hydrogen fusing stars are found. The x-axis of temperature is inverted because it was originally plotted as least to most absorption. Like most things in astronomy: if at first it doesn't make sense, it's a historical thing.

of light received from the star). Hertzsprung plotted this on a graph, showing how the two were correlated. By 1913, American astronomer Henry Russell had collated yet more distance measures to stars, allowing him to calculate more absolute brightnesses and revise Hertzsprung's diagram to yet again show off the correlation between brightness and absorption-line strength. The brightness obviously had something to do with the amount of absorption in the spectrum of stars, but what was the link?

For that, we come back to the work of Cecilia Payne-Gaposchkin (who, if you remember from the previous chapter, showed in her 1925 PhD thesis that the Sun was made mostly of hydrogen). Edward Pickering's inclusion of women computers at Harvard College Observatory, allowing them to publish their work under their own name (uncommon at the time), paved the way for many more women in astronomy. After Pickering died in 1919, the new director of the Harvard College Observatory was American astronomer Harlow Shapley. Shapley established a graduate program in astronomy for women at the observatory in conjunction with the nearby Radcliffe College for women.

Payne-Gaposchkin wasn't hired as a computer, but taken on as a graduate student and subsequently became the first person to be awarded a PhD in astronomy from Radcliffe College at Harvard University.[16] What Payne figured out

---

16 In 1956, Payne-Gaposchkin also became the first woman to be given the title of professor at Harvard, eventually becoming the Chair of the Department of Astronomy. In doing so, she also became the first woman to head a department at Harvard. She supervised many of her own

during her PhD was how the classes of stars (A, B, F, G, K, M and O) were related to their temperature. She had read the work of Indian physicist Meghnad Saha, a professor at Allahabad University in Uttar Pradesh, who was studying the behaviour of gases at high temperatures. Saha used the ideas of quantum mechanics, i.e. how tiny particles behave, to work out what happened to atoms at incredibly high temperatures and pressures. He realised that the higher the temperature or pressure, the more ionised a gas became. The more ionised a gas, the more electrons are released from their orbits around the centres of atoms, giving free-roaming negative electrons and positive nuclei. He wrote this all down in a nice neat equation known as the Saha equation.[17]

Other physicists, such as British astronomer Ralph Fowler, realised the implications of Saha's work; that this would cause different amounts of absorption in the spectra of stars. Too cold and there wouldn't be enough energy to boost electrons up to higher orbits, and so there would be less absorption of light by the electrons. Too hot and there would be so much ionisation that there would no longer be any electrons left in orbits around atoms to steal away the light, and so there would again be less absorption of light. There should then be a sweet spot where the most absorption of light happens by

---

graduate students in her time, including Frank Drake of Drake-equation fame, which attempts to estimate how many other advanced civilizations there might be in the Milky Way.

17 I feel like this is every physicist's dream – to come up with a brand new equation and have it named after them. Either that or a very specific graph.

the electrons, at that perfect Goldilocks temperature, giving many gaps in the spectrum of a star.

It was Celia Payne-Gaposchkin who took these ideas further and demonstrated that Annie Jump Cannon's classification system could be ordered as O-B-A-F-G-K-M from the hottest to the coolest, with the most absorption happening in A stars at that Goldilocks temperature of not too cold and not too hot. Having realised that the amount of absorption was due to temperature, and not the amount of any particular element, she showed that the Sun actually contained a million times more hydrogen than anything else. Her work was published in 1925, but she was dissuaded from making such a bold claim by her thesis examiner Henry Russell, since it went against the thinking of the time that the Earth and the Sun were made of a similar amount and mix of elements. In 1929, Russell, independently, with a different method, determined that the Sun was mostly made of hydrogen, and although he credits Payne-Gaposchkin's earlier work, the credit for the discovery is often mistakenly attributed to him.

Thanks to Payne-Gaposchkin's insight, we now understand how stars shine, the correlation of brightness with the absorption strength and the classification of stars. It's a simple classification system that is still taught to budding astronomers the world over with the handy mnemonic: 'Oh Be A Fine Guy/Girl Kiss Me'. It is known as the Harvard Classification Scheme, rather than the perhaps more fitting Cannon Classification Scheme, with many students learning about the scheme but not the women behind it.

So, because the absorption strength in a star's spectrum is

determined by the temperature of the star, the fundamental relation is that the temperature of a star is correlated with its absolute brightness, something that's now known as the Hertzsprung–Russell diagram. The hotter the star, the more light and, crucially, the more energetic the light will be that shines out. The temperature of the Sun, on average, is 5,778K (kelvin),[18] meaning that it emits the most light at a wavelength of around 500 nm (nanometres, or 0.0000005 m), which is greenish in colour. There's a similar amount of red and blue light being emitted, close enough for it all to mix together to give white light, which is why the Sun doesn't actually look green. Betelgeuse, with its reddish colour, is cooler at 3,600K, and Sirius with its blueish colour is hotter at 9,940K.

But again, *why* are brightness and temperature correlated in stars? The last piece of the puzzle in understanding stars is their mass. Edward Pickering, while driving the cataloguing of all those stars at Harvard College Observatory, was himself studying binary stars; pairs of stars that orbit each other. This allowed him to work out how heavy stars of different spectral types were. The heaviest are the O stars, and the lightest are the M stars. Essentially, the more massive the star, the brighter and hotter it will be.

This makes sense if we think about stars, like Lord Kelvin did, as a constant balance between the crush of gravity inwards and energy from nuclear fusion pushing outwards. The most

---

18 5,500°C (or 9,332°F if you must) to use the non-scientific unit of temperature. To convert between kelvin and celsius, just subtract 273.15 from the temperature in kelvin.

massive stars will exert the greatest crush of gravity inwards, heating up the inside of the star to much larger temperatures than in smaller stars. To resist that bigger force of gravity pushing inwards, more massive stars need a bigger force pushing outwards: they need to burn more fuel each second so they don't collapse under their own gravity. This is why they're brighter – they're fighting harder all the time against their own much stronger gravity. So much so that even though more massive stars are made of far more hydrogen than our Sun, the rate at which they have to fuse that hydrogen means that they live much shorter lives. An O star can be ninety times heavier than the Sun, but live for only a million years (10,000 times less than the Sun's 10 billion years). Bigger stars live fast and die young.

During their lifetime of happily fusing hydrogen into helium, stars are found on what's called the 'main sequence' of the Hertzsprung–Russell diagram: that main correlation of brightness and temperature. But when stars start to run low on hydrogen fuel, they begin to stray off this correlation, cooling down and changing to a redder colour but somehow staying at the same brightness. They do this by swelling up to a huge size, and we class these stars as 'giants' (or perhaps even 'supergiants' if they're particularly large). If you find a big cluster of stars that have all formed at the same time, you can tell how old they are since the brightest O stars will already have died and will be missing from the Hertzsprung–Russell diagram, and there'll be a point where the main sequence switches back on itself to give a large number of giant stars.

Swelling up to these giant sizes is a star's way of delaying the inevitable. For example, when the Sun starts to run out of fuel in 5 billion years or so it will take a very winding path through the Hertzsprung–Russell diagram, swelling up to a red giant and then eventually down to the white dwarf section (at high temperature but low brightness), after losing its outer layers to space. But why does it do this? What are stars doing when they swell to delay the inevitable?

Once astronomers had finally put all the pieces together in 1929, and worked out that the process fuelling the Sun and all the stars in the sky was hydrogen fusion into helium, the work could truly begin to understand how this actually happened. How do you physically get four hydrogen atoms to come together and fuse to make helium? It was in 1939 that German-American nuclear physicist Hans Bethe worked out the way that fusion was actually happening in stars.[19] George Gamow (who worked out the maths for the probability that two hydrogen atoms could overcome the repulsion between them, and realised it was small but not zero) had

---

19 Bethe's mother was Jewish, and in 1933 Bethe found himself dismissed from his research post at the University of Tübingen, due to the newly elected Nazi party's anti-Semitic and racist Law for the Restoration of the Professional Civil Service. After a short stint at the University of Manchester in the UK, in 1935 he moved to the US permanently to a professorship at Cornell University. During the Second World War he then found his nuclear physics knowledge had earned him a position as head of the theoretical division at Los Alamos laboratory, developing the first atomic bombs, such as the one dropped on Nagasaki in 1945. In later life he campaigned alongside Albert Einstein against nuclear testing and the nuclear arms race.

previously suggested a chain reaction of hydrogen atoms merging; first with two fusing to produce heavy hydrogen, also known as deuterium. Deuterium has one proton in its nucleus, like normal hydrogen, plus a neutron, making it slightly heavier.[20] It's the number of protons that determine what element an atom is; the number of neutrons just determine how heavy the atom is. Normally, atoms will have an equal number of neutrons and protons (except hydrogen, which has no neutrons normally)[21], we call these atoms with a different number of neutrons than normal, like deuterium, isotopes. In the chain reaction, the heavy hydrogen then fuses with another hydrogen atom to make light helium (helium-3), which then finally fuses with another hydrogen atom to make helium.

But Bethe wasn't convinced by this proton chain reaction; what about the heavier elements, like carbon, that we knew were also part of the Sun and stars? How did these get made and how did they influence the nuclear reactions going on in stars? Bethe realised that the presence of carbon could actually act as a catalyst for nuclear reactions, at least when stars were hot enough. Stars could cycle through combining hydrogen with carbon, nitrogen and oxygen to finally make some helium in the end. The cycle goes like this:

---

20 If you can't remember or don't know what a proton, a neutron or an electron is, don't worry, we'll get to that in the next chapter.

21 Also excepting ever heavier elements, that become unstable without a whole load of extra neutrons to hold them together against radioactively decaying into lighter elements.

(i)   carbon fuses with hydrogen (#1) to make light nitrogen

(ii)   light nitrogen decays into heavy carbon

(iii)   heavy carbon fuses with hydrogen (#2) to make nitrogen

(iv)   nitrogen fuses with hydrogen (#3) to make light oxygen

(v)   light oxygen decays into heavy nitrogen

(vi)   heavy nitrogen fuses with a hydrogen atom (#4) and splits to give carbon plus a helium atom

In this cycle, we started with carbon and ended with carbon, used four hydrogen atoms along the way and made some helium. It's known as the CNO-cycle (the carbon-nitrogen-oxygen cycle).

Bethe calculated that at hotter temperatures, this process is much more efficient than the proton-proton chain reaction; it's much more likely to get hydrogen to fuse with either carbon or nitrogen than with itself. Bethe published his work in 1940, and in 1967 he won the Nobel Prize in Physics – he'd cracked exactly how the stars were powered.[22] But what it didn't answer was how did carbon, nitrogen and oxygen get made in the first place? Hydrogen is the simplest element, with just one proton in its nucleus, a basic building block of the Universe. Hydrogen is the most abundant element in the Universe, so there must be some other process converting it to things even heavier than helium.

---

22 We now know that stars with around the Sun's mass or less are actually dominated by the proton-proton chain reaction, and only those stars heavier than the Sun power themselves using the CNO-cycle.

Bethe never considered this problem of heavy element creation, and it would be a few more years – 1946 – until British astronomer Fred Hoyle did. Hoyle was a lecturer at St John's College at the University of Cambridge, and his ideas on heavy element production helped make him a household name[23] and eventually led to him becoming the first ever director of the Institute of Theoretical Astronomy in Cambridge. Hoyle suggested that when stars run out of fuel to burn and they no longer have any energy pushing outwards against the crush of gravity inwards, they start to collapse under gravity. This crushing of matter down would increase the temperature inside the star to millions of degrees, causing the hydrogen and helium nuclei created in normal fusion to fuse together to make all the elements across the periodic table in roughly equal abundances.

The problem with this idea is that those elements then become trapped inside the collapsed star, never to see the light of day. But we know that those elements have to be somehow dispersed across the Universe, to give us the ingredients to cook up the Solar System. So Hoyle revised his theory, thinking about this strange giant phase that stars go through when they run out of hydrogen fuel. It's only hot

---

23 He also vehemently argued against the Big Bang theory for the origin of the Universe. He even coined the term 'Big Bang' on a BBC radio programme as a visual description of the theory for the listening British public. Instead, Hoyle insisted that the Universe had always existed and would continue to exist in a steady, unchanging state. He was eventually proven wrong and the Big Bang theory, which he named, came out on top.

enough in the inner core of a star for fusion, so only about 5 per cent of the hydrogen in a star actually gets converted to helium over its lifetime (as Arthur Eddington himself suggested). When a massive star first runs out of hydrogen fuel, the whole thing starts to collapse under gravity, with the outer atmosphere of hydrogen crushing down on the core, now made entirely from helium.

As the star collapses under gravity, the hydrogen closest to the core becomes hot enough to fuse into helium again, and starts heating up the helium core and the hydrogen atmosphere around it. The core contracts some more, getting ever hotter, and the only thing the star can do to balance this out is to swell its outer atmosphere of hydrogen outwards to become really diffuse. It becomes a giant, or if it's a really massive star, a supergiant (the outer layers of the star cool as they get more diffuse, which is why these giant stars then look red).

Fusion continues in a layer around the core, until the core becomes hot enough to start fusing helium into carbon. Eventually, the hydrogen around the core runs out and the star starts to collapse again, before it becomes hot enough in another layer to kick-start hydrogen fusion again. The previous layer that was fusing is now pure helium, and it starts to fuse that into carbon, while the carbon that was made in the core starts to fuse into oxygen. This process keeps on repeating until you're left with a star that resembles an onion, with layer on layer of heavier elements made by fusion triggered by ever increasing temperatures as the star tries to prevent its inevitable collapse.

The star will continue this constant fusing of ever-heavier elements in its core until silicon atoms fuse to give iron. Iron is the death sentence for stars. Iron can fuse together to give heavier elements, but you need to put more energy in than you get out, so it can't be used as a fuel. At this point, when the star contracts again, there's no extra layer made and there's no longer any fusion process that can resist the crush of gravity inwards. The lighter elements on the outskirts of the star collapse inwards, briefly exponentially increasing the temperature and producing a huge burst of light that can be seen across galaxies and beyond, before bouncing back off the heavier elements in the core to get flung outwards into space. We call this collapse and rebound a supernova.[24]

Hoyle published this hypothesis of the onion-like death of stars in 1954, and in 1957 teamed up with three other scientists – American physicist William Fowler, British astronomer Geoffrey Burbidge and British-American astronomer Margaret Burbidge – to write one of the most influential research papers in all of astrophysics: 'Synthesis of the Elements in Stars'. It's known as the B^2FH paper (from the initials of its authors), and is essentially a review that pulled together all the work

---

24 Note that the Sun won't ever do this, as it's not massive enough. In 5 billion years or so it will swell into a red giant, swallowing up the Earth and perhaps even Mars, but won't reach quite the same level of onion-like status as more massive stars. It's not massive enough for gravity to apply enough force to the core to trigger fusion of carbon and oxygen into heavier elements. At that point, the core is so hot it will push back the outer layers of the Sun's red giant atmosphere in more of a fizzle than a spectacular supernova.

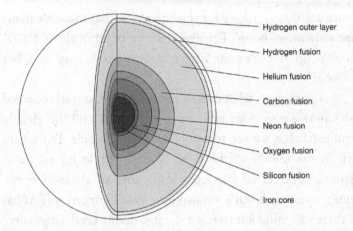

Hydrogen outer layer

Hydrogen fusion

Helium fusion

Carbon fusion

Neon fusion

Oxygen fusion

Silicon fusion

Iron core

The onion-like structure of a supergiant star nearing the end of its life.

that had been done into the production of the heavy elements in fusion (by nuclear physicists), observations of the ratio of the amounts of heavy elements in stars (by astronomers) and Hoyle's ideas on the onion-death scenario of stars. It identified the nuclear reactions that would occur in each of the layers of the dying star, predicted the amount of each element that would be formed and showed how this matched the amounts measured in astronomical observations of the spectra of stars. It tied up fifty years of research with a nice, neat little bow.

The B^{2}FH paper wasn't just influential in the field of astrophysics, it captured the attention of the wider public too. If stars are the great forges of the Universe, with all elements made within them and ejected back into the Universe, then that means me, you, the entire Earth even, are all made of 'stardust'. It sounds very poetic, but my favourite, and I think more accurate, analogy for this process is that these elements are 'supernova poop'. I realise that 'we're all made of supernova poop' doesn't quite have the same poetic ring to it, but I like it.

A supernova is what caused the bright 'guest star' recorded by Chinese astronomers in 1054 and left behind the ghostly remnants that we see today in the Crab Nebula. But what's left in the middle of the Crab Nebula, producing all those gamma rays? What happens to the core of a star after the outer layers of a star's atmosphere have bounced off? What if there's nothing left to resist the inexorable crush of gravity?

We get a black hole.

3

Mountains high enough to keep me from getting to you

If there's one thing I could change in the entirety of physics it would be the name for black holes. 'But what's in a name?' you, or Juliet, might ask. *A lot.* While Tolkein might have argued that 'Cellar Door' were the two most beautiful words in the English language, I'd argue that no two words have ever caused more misunderstanding and misconceptions than 'black hole'. *Black hole* gives people visions of a deep dark well you can fall down, a sink plug hole, or even a cosmic whirlpool stealing away spaceships the same way sailors on the sea have been caught unaware.

Perhaps the most concerning thing of all is that the term black hole leads people to believe that black holes are the absence of something. That they are negative space. Something that takes away. Well, let me be the one to tell you that a black hole is the furthest thing from a hole you can get. A black hole isn't the absence of something, it's the presence of *everything*; matter in its densest possible form. I like to think of them more like mountains of matter than holes in the ground.

So where did this idea of a 'hole' come from? Well, in part we've got Einstein's theory of general relativity to blame. General relativity is, first and foremost, a theory of gravity

– it tells us how objects in space influence other objects and the paths they will take, either in orbit or with a quick deflection. Now you're probably thinking, *isn't that what that guy Newton did when the apple fell on his head?* Technically, yes. As told by many of his contemporaries, in the 1660s British physicist and mathematician Isaac Newton was inspired to think about what force causes things to fall after seeing an apple fall to the ground in his garden in Lincolnshire. He questioned why the apple always fell straight downwards and never varied by falling diagonally, or even upwards, figuring that the apple must therefore always be attracted to the very centre of the Earth. His notebooks from the time show how he puzzled over this idea for many years, wondering whether the force exerted by the Earth extended beyond its surface, perhaps even keeping the Moon in orbit.

It took Newton nearly two decades before he published arguably his most famous work, *Principia*, in 1687, in which he laid out his famous three laws of motion. The first: any object at rest will stay at rest, and any object in motion will stay in motion unless another force acts to slow it down. The second: the force applied to an object will be equal to its mass multiplied by its acceleration (usually people remember this from their high-school days as $F = ma$, after having it drilled into them repeatedly). The third: every action has an equal, opposite reaction[25] – which essentially means if you pull on something it will pull back.

---

25 'Thanks to Hamilton, our cabinet's fractured into factions' – that one's for my fellow *Hamilton* fans.

But Newton didn't stop there. He also defined his universal law of gravitation, which states that every single particle in the Universe attracts every other particle with a force that depends on how massive each one is, and dissipates as they get further away from each other (by their distance squared, so it weakens quickly). So, right now, you are attracted by gravity to this book you are holding and the book to you, but because you and the book are not, astrophysically speaking, that heavy, you barely even feel that pull (it's a force of about 0.000000005 N; the force your back teeth generate when you chew is 1,000 N).

What Newton was suggesting in *Principia* was that there was an invisible force acting over great distances across the entire Universe. It was an idea that was met with huge scepticism by many scientists and philosophers at the time, who accused him of being drawn in by 'occult' ideas; they thought Newton was a crackpot. I like to remind people how you can't see magnetic forces either, but you can still feel the magnetic attraction between two magnets. The effects of magnetism had been known since antiquity, and in 1600 British philosopher William Gilbert had published work outlining how Earth itself was a giant magnet. So the appreciation of invisible forces was already there in the scientific community, but perhaps in the heat of the moment Newton didn't think of this microphone-dropping clap-back.

So while Newton's work in *Principia* gave people a framework for describing gravity, and eventually propelled Newton to international scientific stardom, what *Principia* didn't do was actually explain what gravity was and what caused it, much to the scientific community's chagrin. It would be over 200 years (but not for lack of trying!) before another, different theory of gravity was proposed that actually explained the cause of gravity: Einstein's theory of general relativity. Although Newton's laws of motion and gravitation were eventually accepted by the scientific community, there was one problem with them. Although they could predict the positions of the planets in the Solar System as they orbited the Sun with great accuracy, for the closest planet to the Sun, Mercury, they always gave a slightly wrong answer.

No one knew quite why that was until way beyond Newton's lifetime, when in 1859 French astronomer Urbain Le Verrier figured it out. Le Verrier was already a well-known and well-loved character in the astronomy community at the time, after he observed oddities in Uranus's orbit in 1846. He predicted that they were caused by a large planet beyond the orbit of Uranus and sent a letter to the Berlin Observatory telling them where to look. That same evening, Neptune was discovered just 1° away from where Le Verrier predicted it would be (to understand how accurate that was, hold out your hand at arm's length in front of the sky; your little finger is about 1° across at that distance from your face).

What do you do after predicting the existence of a planet in the Solar System that no one knew existed? Well, Le Verrier turned to predicting the motions and positions of all the planets in the Solar System, to ensure that nothing else had been missed. A mammoth task and one that kept him busy for the rest of his life. In that pursuit he studied the orbit of Mercury by observing its position for many years, and in 1859

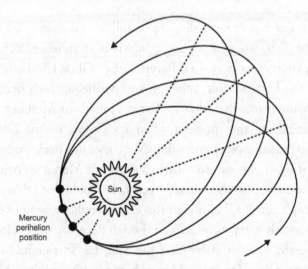

The precession of the perihelion of Mercury. The effect is exaggerated here to show the 'spirograph' shape the precession eventually makes over many millennia.

published his data; a huge long list of the position of Mercury over a number of years. He noticed that what was happening to throw off his (and other's) predictions for Mercury's position was that its perihelion was 'precessing'.

Planets don't orbit the Sun in perfect circles. Instead, they orbit in ellipses, an oval-like shape described by two numbers: its furthest position from the centre (around the Sun this is known as aphelion; *ap-* away from, and *-helion*, from the Greek *helios*, meaning Sun) and the closest position to the centre (known as the *perihelion*).[26] For example, on 5 January every

_____

26 A circle is just a very special case of an ellipse where the furthest and closest positions are equal.

year the Earth is at perihelion 147.1 million kilometres away from the Sun, whereas on the 5 July it is at aphelion 152.1 million kilometres away – a difference of 5 million kilometres!

For the Earth's orbit, aphelion and perihelion each occur in the same place. But what Le Verrier found for Mercury is that perihelion, the moment when it's closest to the Sun, wasn't in the same place every time Mercury came back around on its orbit. If you were actually to draw out Mercury's orbit over a number of the planet's years, it would look like a Spirograph pattern,[27] although the effect wouldn't be noticeable over only a few orbits. Even though Mercury only takes eighty-eight days to go around the Sun, Le Verrier had to wait many orbits for this effect to become apparent to able to detect it.

In one sense, what was happening to Mercury's orbit wasn't that much of a surprise, as Newton himself had predicted it. When there's a smaller object quite close to a massive object with other objects orbiting it, the smaller object is perturbed slightly by all the other objects in the system. So the main reason why Mercury's orbit precesses is because it's not just interacting with the Sun, but it's also feeling the pull of all the other seven planets (plus all the dwarf planets, comets and asteroids littering the Solar System) that are also orbiting the Sun. But Le Verrier was the first to point out that if you use the equations in Newton's theory of gravity to predict how

---

27 Spirograph was one of my favourite games as kid. I obsessively produced every variety of Spirograph with every different-smelling and -coloured gel pen I could get my hands on.

much Mercury's orbit should precess per century, it's a smaller value than you observe.

Before declaring that there must be something wrong with a law of gravity that had been accepted for over 170 years, Le Verrier considered other explanations for the discrepancy. Including that the Sun is not perfectly round but is instead an oblate spheroid, meaning it's a bit squashed at the poles. The same is true for Earth, and especially for Saturn because it's rotating so quickly; matter at the equator bulges out a bit, in the same way you feel a force pushing you off a merry-go-round. It did turn out that that the Sun's shape plays a small role in how much Mercury's orbit precesses, but it still wasn't enough to account for the discrepancy. So Le Verrier also suggested that there could be another planet inside the orbit of Mercury, orbiting the Sun much closer.

At the time, this extra planet was the most favoured hypothesis to explain the discrepancy, partly because, just thirteen years before Le Verrier had posited it, he had predicted the existence of Neptune due to the effect on Uranus's orbit. So, as odd as the idea of an extra planet between the Sun and Mercury sounds to you and me, back then it wasn't such an outlandish idea. Neptune had only just been discovered and there was a general feeling that there must be something else out there. So finding this hypothetical planet between the Sun and Mercury (dubbed Vulcan after the Roman god of volcanoes and fire and forges) became the focus of many astronomers during the rest of the nineteenth century.

The desire to be the person responsible for the discovery led to a lot of false claims, including people who were adamant that they'd observed a planet very close to the Sun during a solar eclipse, in a position where no known star was thought to be (in the background), despite no one else observing it during the same solar eclipse. All of these false claims gave rise to different descriptions of the properties of Vulcan and its orbit; if all the claims had agreed on the properties then perhaps the idea of a new planet inside Mercury's orbit would have been quite convincing, but it became clear pretty quickly that this hypothetical planet was just that, hypothetical, and couldn't explain the strange precession of Mercury's perihelion.

So, with all other options exhausted, the only explanation was that Newton's theory of gravity wasn't quite right. This is where Einstein comes in. In the first decade of the twentieth century, Einstein announced his theory of special relativity to the world, which described what happened to your perception of time and space when you travelled close to the speed of light. It introduced the ideas of time dilation – the faster you travel the less time passes from your perspective – and length contraction – the faster you travel your length contracts in the direction you travel. Like most revolutionary theories, this was incredibly controversial and it left many unanswered questions. Trying to tie up all the loose ends, Einstein ended up coming up with a new way of explaining gravity: as the curvature of space itself. Massive objects curve the space around them and then anything travelling along that space, whether a planet or light, would follow a curved path. People often picture this as a sheet stretched taut, or a trampoline, with a basketball placed in the middle. If you then roll a ping pong ball along that surface it will follow a curved path, even if you set it off on a straight one. While that's a great analogy, it doesn't help us visualise the curvature of space in three dimensions, something the human brain can't quite wrap its head around.

Einstein published his theory of general relativity in a series of papers between 1907 and 1915, and in them he proposed the equations that essentially describe the curvature of space that massive objects cause. It was a generic equation that could be applied to many different scenarios depending on different masses and, crucially, the different speeds the objects were travelling at: everyday speeds or close to the speed of light. Einstein found that when he applied his theory of general relativity to the problem of the Solar System, his equations reduced down to match Newton's equations when objects weren't moving at speeds close to the speed of light or close to very massive objects. So it wasn't that Newton's equations were wrong necessarily, it was just that they were a generalisation for a special case. Mercury, though, is close to the massive object of the Sun, and so Einstein's equation for Mercury's orbit was ever so slightly different from Newton's. What Einstein did was work out how much of an effect this difference in the equation would have on the predicted position of a planet and in particular how much precession of Mercury's perihelion was expected. He found that it was the same value measured by Le Verrier and used this as evidence for his newly proposed theory of gravity. He suggested two other phenomena that would also provide evidence for his new theory: massive objects should cause redshift of light (this stretching of the wavelength of light, 'gravitational redshift', was finally confirmed in 1954) and also the bending of light by massive objects.

In Einstein's lifetime it was only possible to detect the latter: the bending of light from distant stars from behind the Sun during a solar eclipse. During an eclipse, it becomes dark enough to see the stars behind the Sun during the day that are usually only visible at night, six months earlier, when the Earth is on the other side of the Sun. You can compare the positions of the stars at night to the positions recorded during a solar eclipse, and see if the apparent positions of the stars change because their light has been deflected by the Sun curving the space around it. To do just that, British astronomers Frank Dyson and Arthur Eddington (who was already very well-known at the time for explaining general relativity to the English-speaking world after normal scientific lines of communication were disrupted during the First World War, but hadn't quite reached his Big Name in Physics status yet with his work on the fuelling of stars) organised two expeditions to observe the solar eclipse of May 1919.[28] One expedition was to the Brazilian town of Sobral, led by Andrew Crommelin and Charles Rundle Davidson from the Royal Greenwich Observatory, and the other expedition was to the West African island of Príncipe, led by Eddington himself and Edwin Cottingham.

Despite some bad weather during the eclipse, Eddington

---

28 The organisation of this expedition also allowed Eddington to avoid conscription into the British army during the First World War at the age of thirty-four. He claimed to be a conscientious objector due to his Quaker beliefs.

An image of the eclipse as observed by Eddington and Cottingham from Príncipe in 1919.

obtained enough images to record the positions of stars and declare that the change in their position matched those predicted by general relativity. The results were announced at a meeting of the Royal Society in November 1919 and by the next day had made headlines all around the world. The most famous of which was the headline from *The New York Times* published on 10 November 1919, which read: 'Lights all askew in the heavens . . . men of science more or less agog . . . nobody need worry.'[29] It made Einstein, as the man who 'corrected' Newton with his new theory of gravity, world famous, although acceptance of general relativity in the wider scientific community took some time.

---

29 I particularly enjoy the fact that part of the headline reassured people that there was nothing to worry about.

First, because one experiment with one measurement is never enough for us scientists. It had to be repeated, but solar eclipses unfortunately don't come along every day and the weather likes to get involved and ruin the party. Second, the understanding of general relativity among other scientists of the time wasn't great. Einstein's articles had been published in German, and not everyone could get an accurate translation in their own language, mainly because translators had to also be intimately familiar with physics and general relativity as well.

One thing Einstein never predicted with general relativity was black holes (it's a common misconception that he did), although a rough draft of the idea of a black hole had been knocking around long before Einstein. In 1783, British clergyman by day, astronomer by night John Michell mused on the idea of objects so massive that light could not escape them, and dubbed them 'dark stars'. He even went as far as saying that if they existed, we could still spot them by their gravitational pull on other visible objects.

It was German physicist and astronomer Karl Schwarzschild who, in 1915, just a few months after general relativity was published, unknowingly found the first mathematical description of a black hole by solving Einstein's equations (more on that later). One possible scenario that Schwarzschild's solutions described was all mass collapsing down to a single point. In this scenario, many terms in the equations became infinite. Even time itself would stop, which led to these objects being referred to as 'frozen stars'. But if we think of this in terms of how Einstein described gravity, as the curvature of space and time, and go back to our analogy of the trampoline, we can imagine how putting an incredibly dense, heavy object on the trampoline would cause a very steep-sided depression. A hole, you might say. Yes, as much as we have Einstein to thank for, we perhaps also have him to grumble at for the idea of a 'hole' in space being planted into people's brains.

Of course, the physicists of the day did not accept that Schwarzschild's solutions were realistic, merely theoretical curiosities. What we now call 'black holes' were referred to as 'gravitationally collapsed stars' or just 'collapsed stars', which is also how prominent Swiss astronomer Fritz Zwicky referred to them in a paper in 1939. But by 1971, Stephen Hawking himself, in his paper 'Gravitationally Collapsed Objects of Very Low Mass', refers to them as 'black holes' in inverted commas. So where had this term come from in the time between the 1940s and the 1970s? What's the etymology of the phrase 'black hole'?

It seems we have famous American physicist Robert H. Dicke to blame for coining the phrase that eventually made its way around astronomy research circles. Unfortunately it's a rather harrowing tale from a sad part of history that seems to have inspired Dicke. At the first Texas Symposium in Dallas, in 1961, attendees reported that in his presentation Dicke repeatedly compared 'gravitationally completely collapsed stars' to the 'black hole of Calcutta'; a small prison cell in the dungeon of Fort William in Kolkata, India that measured just 4.30 × 5.50 metres (14 × 18 feet; about the size of three double beds).

Fort William was built to defend the British East India

Company's trade in Kolkata. However, the leader of the region, the Nawab of Bengal, Siraj ud-Daulah, ordered the construction be halted. The British carried on anyway, and in retaliation Siraj ud-Daulah's forces laid siege to the fort. The majority of the British troops were ordered to abandon their posts and escape, except for 146 soldiers who were left behind as a last defence. The fort fell in June 1756 and the surviving British soldiers were all imprisoned in the 'black hole'. The conditions were so cramped, with so many in such a small space, that overnight people died from suffocation and heat exhaustion. Reports vary on the number of lives lost, but historians estimate that sixty-four people were imprisoned and only twenty-one survived the night. There is a memorial at St John's Church in Kolkata, which was erected in 1901 to those 'who perished in the Black Hole prison of old Fort William'.

It was this historical event – of people being crushed in the prison – that rather morbidly led Dicke to use the term for when matter has been crushed and a star collapsed down due to gravity. One of his colleagues who picked up on the phrase was American physicist Hong-Yee Chiu (who is credited with inventing the word 'quasar' – a portmanteau of 'quasi-stellar object'). He inspired science journalist Ann Ewing to write an article called '"Black Holes" in Space' for the magazine *Science News Letter* in 1964, which marks the first time the term was ever used in print.

It was John Wheeler who is credited with truly popularising the name, though, turning the term from analogy into actual scientific jargon.[30] In 1968, he was giving a presentation at the NASA Goddard Institute in New York about his recent work studying 'gravitationally completely collapsed objects' when he jokingly complained that the term was too long and far too inconvenient to repeat all the time. According to Wheeler in his autobiography, someone in the audience at that point suggested 'How about black hole?', and he thought the term was perfect for its brevity and 'advertising value'. He then adopted the term whole-heartedly, using it in an 1968 article for the *American Scientist* journal. The term quickly entered the scientific lexicon, with German astrophysicist Peter Kafka the first to use it in a scientific research article in 1969, with the likes of Stephen Hawking following suit by 1971. The term 'black hole' had stuck; much to my later annoyance.

I guess I should be grateful that the modern trope of shortening everything in astronomy to an acronym hadn't quite gained traction in the 1960s, otherwise I'd probably be telling everyone that I study 'GCCOs' (gravitationally completely collapsed objects, d'uh). But what *would* I have named black holes instead if I'd had the chance? If I'd been there in the 1960s and had the same influence as Wheeler to dub these most spectacular of objects?

Honestly, I'm not sure, but if I had to choose, I think John Michell's 'dark stars' is my favourite, and would cause less

---

30 Similar to how Hoyle's 'Big Bang' analogy eventually made it into the scientific lexicon.

confusion over what black holes really are.[31] Or perhaps 'mountain' might be a better word to describe the nature of black holes – because it's not like the stuff that does 'fall' into a black hole just disappears. In fact, the material piles up and up, so much so that in some cases there can be over a trillion times the mass of the Sun squashed into a black hole. That is a literal mountain of matter. Just mountains that you can't directly see because not even light can escape. I don't want to have be the one to break it to Tammi Terrell and Marvin Gaye, but it turns out there are mountains high enough to keep me from getting to you.

---

31 But here's a secret for you to keep until Chapter 7: they're not *technically* dark.

4

Why black holes are 'black'

To understand why there are mountains high enough to keep me from getting to you, essentially why black holes are even 'black' in the first place, we first have to understand light itself. The history of our understanding of light is fascinating. Early philosophers, such as Euclid and Ptolemy, thought that our eyes themselves generated light, which then allowed us to see the world around us. Upon hearing this logic, Heron of Alexandria declared that if that was the case then the speed of light must be infinite and instantaneous, because as we open our eyes we see the light from stars at great distances instantly. We're in the privileged position of hindsight here, knowing that our eyes don't produce light but instead use our rod and cone cells to *detect* light, so we know this argument was flawed from the start. Yet the idea that the speed of light was infinite was still knocking around in the seventeenth century, with both Johannes Kepler and René Descartes (two giants of mathematics and astronomy) supporting the idea.

It was Galileo Galilei[32] (of discovering Jupiter's moons with

---

32 I love that Galileo is widely known by only his first name (a mononymous person). It puts him in such interesting company, with the

his telescope fame in 1638) who was the first to actually try and measure the speed of light. His experiment consisted of two people on hilltops separated by one mile; one with a covered lantern which they would uncover and then record the exact time. The person on the other hilltop would then record the time at which they saw the light from the other person's lantern. In Galileo's experiment, the two people recorded the exact same time for the lantern being uncovered, and many philosophers at the time took that to mean that the speed of light must be infinite. But Galileo himself pointed out that the results of the experiment could also mean that light travelled too fast for a person a mile away to record a difference. He was right: it takes light just 0.000005 seconds to travel one mile, whereas an average human reaction time (i.e. your eyes detect light, send signals to your brain, your brain makes a decision and tells your muscles to react) is about 0.25 seconds.[33] To put that into context, it takes light less than that, about 0.133 seconds, to circumnavigate the Earth around the equator. So early scientists never really had a hope of measuring the speed of light on Earth, because the distances they used were just too short.

Having failed to put a number on the speed of light, Galileo gave up on that idea and turned his attention to another problem entirely: navigation. Galileo was living through the era of the first regular transatlantic voyages, and knowledge

---

likes of Hercules, Boudicca, Michelangelo, Madonna and Beyoncé. Now there's a dinner party I'd like to be at.

33 You can test your reaction times with various different websites online. I just tested mine (I procrastinate a lot when I write) and it averaged out over five tries to about 0.263 seconds.

of how far north–south and east–west a ship was could make the difference between life and death.

Figuring out how far north or south you are – your latitude – is fairly easy. On the equator, the Sun at noon is directly overhead (at least on the equinoxes anyway, when the Earth isn't tilted towards the Sun), but if you move further north or south, then the highest point that the Sun reaches in the sky drops. The angle it drops by is how far above the equator you've gone around the Earth. If only it were that simple all year round though, because the Earth's axis is also tilted by 23°, which gives us our seasons. So there's a little added complication, but essentially if you know roughly what time of year it is and you can measure the altitude of the Sun above the horizon at noon, then you can figure out how far north or south of the equator you are. Two fairly easy things to know and keep track of.

But what about figuring out your longitude – how far east or west you are? Today, a friendly airline pilot will usually tell you the local time as you land so that you can adjust your watch accordingly. Or by the magic of modern technology, your phone will automatically change to the right time zone. For example, landing in New York after flying from London, you'll set your watch back by five hours after changing longitude by 75° to the west (that's about 20 per cent of the way around the full 360° of the Earth, so about 20 per cent of a twenty-four hour day, which is 4.8 hours). Knowing your 'time zone' is therefore the key thing to working out your longitude, and in the seventeenth century, governments, kings and queens were all too aware of this.

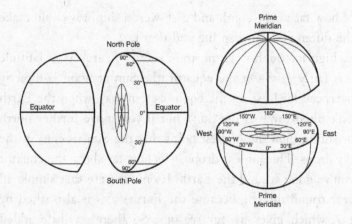

Latitude (left) and longitude (right) on Earth.

The problem was that they didn't have any way of knowing the time in two separate places at once. Ideally, you'd set a watch to the time in Lisbon as you set off on your transatlantic voyage and watch how the Sun reached its peak at noon later and later every day according to that watch. Knowing the time difference between noon where you are and noon at the place you left gives you your 'time zone', and your longitude. But accurate mechanical clocks weren't invented until the eighteenth century; in the seventeenth century sundials were the only means of telling the time, which only tell your local time with respect to the Sun, and not the time of the destination you left. Rewards were set up with big-money prizes by everyone from the British government to King Phillip III of Spain, in the hope that someone could crack the problem and work out a way to know the local time at sea.

It was Galileo and his precious moons of Jupiter that

provided the first glimmer of hope. Just like our own Moon, which orbits the Earth every twenty-eight days like clockwork, Jupiter's moons orbit with the same cosmic clock precision. The four largest 'Galilean' moons of Jupiter can be seen with a simple pair of modern binoculars (you need about 15x magnification) and it was these that Galileo meticulously observed, recording the time it took for each to make one orbit around Jupiter. A handy marker point was when each moon would disappear behind Jupiter and reappear on the other side, i.e. they were eclipsed by Jupiter from our perspective. These eclipses were incredibly predictable; the innermost moon, Io, orbits Jupiter every forty-two hours (just shy of two Earth days), and so giant spreadsheets could be made with predictions of the exact time in, say, Paris that Jupiter would eclipse Io.

Galileo's idea was that if you could observe the time of the eclipse from wherever you were at sea and compare it to the predicted time of the eclipse in Paris's time zone, then you'd be able to work out your longitude. He pitched this to the king of Spain in around 1616, and I can only imagine that the excitement was palpable. However, there were two problems. First, Galileo's predictions weren't accurate enough. If your estimate of how long it takes Io to orbit Jupiter is off by a few minutes, that error adds up very quickly over a few weeks, never mind the months that it would take to cross the Atlantic. Second, Galileo, ever the scientist and not a sailor, didn't foresee the impracticalities of trying to observe Jupiter through a telescope on a ship that is at the mercy of the rolling sea. Unsurprisingly, the king of Spain was unwilling to award Galileo the cash prize.

So while Galileo's method was quickly dismissed as impractical for sea voyages, it could still work on land, where map-makers were also clamouring for a more accurate way of determining longitude. They just needed more accurate predictions for the timings of the eclipses. In 1676, Ole Rømer and Giovanni Cassini swooped in to save the day. Rømer[34] was a Danish astronomer working at the Observatoire de Paris as Cassini's assistant, where they picked up Galileo's work and determined the timings between eclipses to a high degree of accuracy. The problem was that the number they measured kept changing from month to month. They noticed that the time between two eclipses got shorter and shorter as Earth travelled in its orbit around the Sun towards Jupiter, and got longer and longer as Earth moved in its orbit away from Jupiter.

The explanation that Cassini came up with was that the light from the eclipse had further to travel as Earth moved away from Jupiter, and therefore the speed of light was not infinite. Cassini announced this explanation to the scientific world in 1676, but was highly sceptical of it himself, and continued to raise other possibilities. Rømer, however, was a big supporter of the idea and set out to prove it by coming up with a way of predicting the eclipse timings for Io based on the relative positions of Earth and Jupiter. He focused on the geometry (rather than actually trying to measure the speed

---

34 Fun fact, Rømer also invented what we'd recognise as the modern thermometer, showing the temperature between the freezing and boiling point of water.

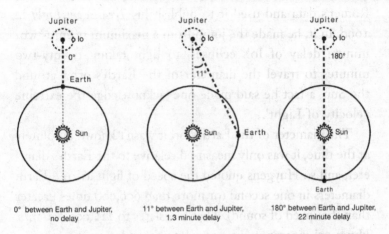

0° between Earth and Jupiter, no delay

11° between Earth and Jupiter, 1.3 minute delay

180° between Earth and Jupiter, 22 minute delay

The delay in the timings of Io's eclipses by Jupiter is due to the relative positions of Jupiter and Earth. If you know the angle between them, you can work out the delay due to the extra distance the light has to travel.

of light) and worked out that there'd be a delay to the eclipse depending on the angle between Jupiter and the Earth. The maximum delay was twenty-two minutes, when Earth and Jupiter were at their furthest distance from each other (an angle of 180°) and that reduced by the fraction of the angle as they got closer together.

It took Rømer eight years of careful observations to work out this time delay, but it allowed for accurate predictions of the eclipse to calculate longitude. This is what Rømer cared about, not measuring the speed of light. While he believed his observations showed that the speed of light was not infinite, he never actually used them to calculate a value. Enter Christiaan Huygens, a Dutch astronomer who took

Rømer's data and used it to publish his *Treatise on Light* in 1690. In it, he made the jump from a maximum twenty-two-minute delay of Io's eclipses, to light taking twenty-two minutes to travel the diameter of the Earth's orbit around the Sun, a fact he said made one 'acknowledge the extreme velocity of Light'.

The diameter of the Earth's orbit wasn't known absolutely at the time, it was only measured relative to the Earth's diameter, and so Huygens quoted the speed of light as 16⅔ Earth diameters in one second (or more than 600,000 times greater than the speed of sound). That translates to 212,000,000 m/s. Huygens' measurement was a bit short (due to the inaccuracies in the relative size of the Earth's orbit to the size of the Earth); the modern value for the speed of light is 299,792,458 m/s,[35] so he was right to within an uncertainty of around 30 per cent. His measurement was a true milestone of scientific history, because it marked the first time that a universal constant (something that is the same across the entire Universe) was measured by humanity.

Of course, Huygens didn't know that at the time, and neither did the multitudes of scientists who came after him who refined that measurement to greater accuracy over the next two centuries. It would take until the early twentieth century, with our old friend Albert Einstein, to understand

35 This is now a definition for the speed of light – we no longer measure it. The speed of light is a universal constant, but the metre is a human construct whose length is completely arbitrary. So we no longer measure the speed of light, but have defined it as 299,792,458 m/s, and instead measure the length of a metre to extreme precision.

why the speed of light was both universal and the finite speed limit of everything in the Universe. It all comes down to his most famous equation ever: $E = mc^2$ ('E equals m c squared'), which if you remember from earlier means that energy and mass are *equivalent* – they are the same thing. But this is actually a simplified version of the full equation, for a specific case when objects aren't moving. If something is moving, the full equation becomes:

$$E^2 = m^2c^4 + pc^2$$

p in this equation is the momentum. Momentum is essentially a measure of how much mass is in how much motion. The more momentum you have, the harder it is to stop your motion. For normal everyday objects, the momentum is mass multiplied by velocity (a speed with a direction, $p = mv$). So, normally, to increase momentum, and therefore the overall energy you have, you need to increase your speed. Which is fine for speeds on Earth: you put an extra bit of energy in and your speed goes up proportionally.

But Einstein's equation above deals with 'relativistic speeds' – those close to the speed of light where strange things start to happen to your perception of time and space. Your momentum at these speeds is once again more complicated than for everyday speeds, so much so that as you approach the speed of light, your momentum no longer goes up proportionally. Instead, your momentum starts to exponentially increase. At 99.99 per cent of the speed of light, the momentum of an object is seventy times more than the value

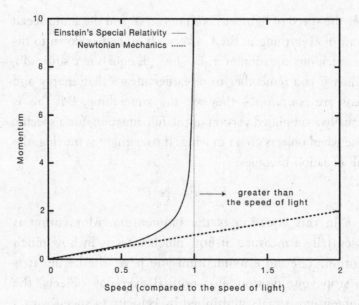

The difference in momentum for Einstein's special relativity
compared to everyday objects explained by Newtonian mechanics.
It shows why you can never go faster than the speed of light
because your momentum and energy tend to infinity.

you'd expect it to be. At the speed of light, an object has an infinite amount of momentum.[36]

This is true not just for momentum, but for all other properties, including the kinetic energy, or movement energy,

---

36 Instead of $p = mv$, there's once again another term: $p = \frac{mv}{\sqrt{1-\frac{v^2}{c^2}}}$. So for everyday speeds, v^2/c^2 ends up as a really small number, so the bottom of that fraction just ends up being 1 and you get back the normal $p = mv$. But for speeds close to the speed of light you end up having to divide by a small number, increasing your momentum. When $v = c$ then you end up dividing by zero to get infinite momentum.

of an object as it travels closer to the speed of light. And as $E = mc^2$ tells us, energy and mass are in principle the same thing. So if the energy of an object shoots up to infinity as it approaches the speed of light, then so too does its mass. So the faster you travel the heavier you get. As you approach the speed of light your mass approaches infinity. There is no number larger than infinity. When you're travelling close to the speed of light, putting in more energy to try and go faster will increase your energy and mass, but not your speed. This is why nothing can go faster than the speed of light and it's why 299,792,458 m/s is the ultimate speed limit across the entire Universe.

This limit on the speed of light is why black holes exist in the first place; it's why they are 'black'. If light speed was infinite, we'd be able to see what a black hole actually looked like, to where all that matter is squished and contained. Instead, light gets trapped in there because the escape velocity of the black hole is greater than the speed of light. All objects with mass in the Universe have escape velocities – the speed you'd need to travel to escape the pull of gravity from that object. Earth has an escape velocity, which is unfortunately much larger than the speed we can jump up at, or throw a ball up at, giving the age old adage: *what goes up must come down*. It's why rockets need to burn an exorbitant amount of fuel to accelerate to a speed where they can fully escape the pull of Earth's gravity, in order to power on out into the Solar System. The escape velocity depends on the mass of the object and how far away you are from the centre of that object; so on the Earth's surface the escape velocity from Earth is around

11.2 km/s (thirty-three times the speed of sound), but on the Moon's surface it's much less, around 2.4 km/s.

For a black hole, there is nothing in the Universe that can travel faster than its escape velocity, not even light itself. It means that we will never observe what black holes truly look like, only their influence on objects around them due to their extreme gravity. But, so long as light doesn't get too close to a black hole, its path through space can instead just get bent to the extreme, like the light from distant stars during a solar eclipse observed by Eddington, but turned up to eleven.

At that point, you can't trust what your eyes are showing you any more, because the black hole has interfered with the light. In 2021, astronomers even detected light from *behind* a black hole. Imagine for a moment that I got into a spacecraft with this book and went around the far side of the Moon and hid where I couldn't see Earth and you couldn't see me. Imagine I cracked open this book to this page and shone a torch on it. Imagine that the Moon was so heavy that the light reflected off this page travelled on a curved path around the Moon and back to Earth so that you could still detect the light from it and read these words. That's what a black hole can do: manipulate light to give you a glimpse of things you shouldn't even be able to see.

5

A teaspoon of neutrons helps the star collapse down!

The recipe for making a black hole is theoretically very simple, yet in practice rather difficult. Essentially, throw enough matter into a small enough space, crush it down and voila! A black hole will result. Now I can't speak for everyone, but my puny noodle arms definitely aren't strong enough to crush matter down in this way, and I imagine neither are yours. I'm sure even veterans of the recipe game like Mary Berry would struggle to follow that one.

Luckily for us,[37] there are processes in the Universe which can follow this recipe with relative ease, thanks to gravity. Annoying as gravity is, keeping us hostage here on Earth, we do also have our very existence to thank it for. In essence, gravity likes to clump things together, whether that's two tiny fundamental particles or two rather large lumps of rock. The force that ruled the early Universe and gave us the first structures out of just tiny atoms of hydrogen, is the same force that turned a random clump of gas on the outskirts of the Milky Way into the Solar System: Earth and all.

At the beginning of the Universe, space, time and the basic

37 I'm sure many of you will argue with my use of the word lucky in this context.

building blocks of matter were formed: protons, neutrons and electrons. Eventually, when the Universe had cooled enough from its hot dense state, those building blocks came together to make atoms, the majority of which were hydrogen atoms. That's pretty much all there was in those days – it's why the early Universe is described as a 'soup of hydrogen', because nothing better describes the boring uniformity of it all than soup. But here's the kicker: technically it wasn't *quite* uniform. In the first fractions of a second of the Universe's life, tiny random quantum flutters made some bits of the Universe slighter denser and some a bit emptier. As the Universe expanded, these tiny quantum flutters grew like ripples on a pond, with more hydrogen forming in some places than others.

Those areas that were already slightly denser, with just a bit more hydrogen, slowly started to clump together and attract yet more hydrogen. And slowly, over a few hundreds of millions of years, enough hydrogen clumped together to become hot and dense enough for hydrogen atoms to fuse together to make a helium atom, and the first stars were born. If this was a recipe, the Universe got all the ingredients out of the cupboard, quantum flutters and gravity took care of the mixing, and finally the first stars started the cooking. When the first stars ran out fuel, supernovae then littered space with the heavier elements – things like carbon, nitrogen, oxygen and iron – polluting pristine hydrogen gas with what astronomers refer to as *dust*. That dusty gas got recycled by gravity to form the next generation of stars, in a cycle of clumping under gravity, fusion and yet more supernova pollution.

Eventually, after there'd been a few generations of stars in one region of the Universe, there was enough dust for gravity to start clumping it together to give solid objects that we might recognise as lumpy asteroids around newly formed stars. If gravity continued to get its way, those lumpy bits of rock kept on clumping together to form planets, moons and entire star systems like our own Solar System. Unfortunately for us, our own Solar System is not destined to become a black hole. The Sun will leave behind a core that's a messy mix of helium, carbon and oxygen that will glow like the dying embers of a fire – something we call a white dwarf.

But what's stopping a white dwarf from collapsing into a black hole? In fact, what's to stop any star as it goes supernova from collapsing into a black hole? With no fusion, surely there's nothing left to stop the endless crush of gravity inwards that has shaped the rest of the Universe around us? To find out why not all stars become black holes, we must once again understand the world of the very small: of atoms, themselves made up of protons, neutrons and electrons.

Figuring out what the building blocks of all things in the Universe are is a question that has been asked by humans since we learnt to ask questions. The basic idea that everything could be made up of tiny particles that are indivisible is a very old one, found in many ancient cultures from India to Greece. These particles were dubbed 'atoms', from the Greek *atomos* meaning 'uncuttable', i.e. these are the basic building blocks of all matter and they are indivisible. There is nothing below an atom.

That idea, that the atom could not be split, pervaded both religious and scientific minds until the late nineteenth century, when a discovery sent people reeling. In 1897, British physicist Joseph John 'J. J.' Thomson was experimenting with something called cathode rays at the Cavendish Laboratory at the University of Cambridge. Cathode rays are generated when two metal rods, one positive and one negatively charged, are placed in a vacuum (a space where all the air molecules have been sucked out). Usually these are in glass tubes, and if you leave a tiny bit of the air in there, you can see a slight glow caused by the cathode rays travelling from the negative to the positive rod. These cathode ray tubes look a bit like a modern neon light sign, and were used throughout the twentieth century in the backs of old-style television sets.

Thomson was trying to figure out what cathode rays were made of. The slight glow must be caused when *something* impacts with molecules in the glass, causing them to give off light. But what something? Thomson decided to try and measure the mass of whatever the cathode ray was made of and was shocked to find that the individual particles were over 1,000 times lighter than a hydrogen atom, the lightest 'indivisible' atom known. What's more, he found that no matter what type of metal rod he used to produce cathode rays, the mass of the particles making it up never changed. The mass was the same no matter the type of atom they came from. He concluded that the only explanation was that the cathode rays were made of a very small negatively charged particle (since they travelled from the negative to the positively charged rod), which were a universal building block in all atoms. They were *subatomic* particles. The atom had been split.

What Thomson had discovered was the electron (although he originally dubbed them corpuscles – there's a name I'm glad didn't stick), and with it he redefined how we think of atoms.[38] No longer were they indivisible, they were made up of yet smaller particles, like electrons, but what else? Atoms were known to be neutral, so Thomson reasoned that there must also be something positively charged that atoms were also made of. In 1904, he proposed what has become known as the 'plum pudding model' of an atom; a sphere of positively charged matter within which the electrons were embedded, like the fruit in a plum pudding.

The plum pudding model, although delicious-sounding with custard, did not stand the test of time, lasting less than a decade before another model usurped it. It was one of Thomson's own protégés, New Zealand physicist Ernest Rutherford,[39] who would find the evidence against the plum pudding model. Rutherford had been working with Thomson in the Cavendish Laboratory in 1897 when Thomson discovered the electron, but Rutherford was distracted by Henri Becquerel's recent (1895) discovery of the strange properties of uranium, and like Marie Curie, set out to investigate further. It was Rutherford who coined the term 'half-life' for radioactive elements, realising that the time it took for half of a

---

38 There's a plaque commemorating Thomson's discovery outside the old Cavendish Laboratory building in Cambridge where the discovery was made. It's on Free School Lane, right in the very centre of town, an unassuming yet quintessential university town side street and well worth a visit.

39 Fun fact: Rutherford's daughter Eileen Mary Rutherford married the physicist Ralph Fowler, who realised the implications of the ionisation of gases being linked to absorption in stars. We'll meet him again later in this chapter.

sample of radioactive material to decay was always the same, thereby giving the geologists the information they needed to figure out how old the Earth was.

In 1907, he moved to the University of Manchester, where he continued to study what was emitted by radioactive elements when they decayed. He had already identified three different types of radiation, which he dubbed alpha, beta and gamma (this is where gamma rays of light get their name from), and showed that when the decay happens an atom spontaneously transforms into another type of atom (another element). It was for this that he won the Nobel Prize in Physics in 1908. Not one to slow down after winning the highest honour there is, Rutherford did his most famous work in the years following his Nobel Prize victory, on the nature of alpha radiation.

Working with German physicist Hans Geiger (of Geiger counter fame – the device for counting radioactive particles), he showed that alpha radiation was made of particles with a charge twice that of a hydrogen atom. Then, working with British physicist Thomas Royds (a local boy at Manchester University, having been born in Oldham), he managed to show that you could make helium using alpha particles; we now know alpha particles are helium atoms with their electrons removed, hence why they are positively charged. To understand this properly, Rutherford wanted to measure the ratio between the charge and mass of alpha particles (this was also how Thomson had managed to discover the nature of the electron). To do this, you set alpha particles moving through a magnetic field and measure how much they are deflected (the greater the charge the greater the deflection, but the heavier the mass the more it will resist deflection). The problem was that the particles kept pinging off molecules of air that got in the way, like a pool break shot scattering balls everywhere, making the measurement unreliable.

Thomson had had the exact same problem when he was measuring the charge-to-mass ratio of an electron, and he solved the problem by doing the entire experiment in a perfect vacuum (i.e. removing all the pesky air in the way). Rutherford didn't think he'd have to do the same thing, because alpha particles were much heavier than electrons (about 4,000 times heavier), and in Thomson's plum pudding model of the atom the sphere of positive charge wasn't concentrated enough to be able to deflect a particle that heavy.

Rutherford decided to investigate this scattering very carefully, with the help once again of Hans Geiger and British-New Zealand physicist Ernst Marsden.[40] Together, they fired alpha particles at thin sheets of gold foil in a vacuum and recorded where the alpha particles ended up. The overwhelming majority went straight through the foil unhindered, but a small fraction were deflected. Most of those alpha particles were deflected by small angles, but again a small fraction of those were deflected so much that they made complete U-turns, coming back towards where they were fired from.

With this new information, in 1911 Rutherford concluded

40 Marsden was born in Britain but lived the majority of his life in New Zealand. Rutherford did the opposite; having been born in New Zealand he lived the majority of his life in the UK.

that the only way to explain what they'd found was if the positive charge in an atom was concentrated in a tiny section right in the centre, orbited by electrons that have a much lower mass. In his model, 99 per cent of the atom was empty space, which allowed the majority of the alpha particles to sail straight through the foil made of gold atoms. Rutherford continued his experiments with atoms and by 1920 had figured out that the hydrogen atom, as the lightest possible atom there was, must have a nucleus made of another basic sub-atomic particle, which he dubbed the proton.

This paradigm shift in the structure of the atom – from indivisible to made up of yet more particles, arranged almost like the Solar System itself – set in motion one of the largest knowledge jumps humanity has experienced. From an understanding of the periodic table and the chemistry underlying everyday reactions, to creating the entire field of quantum mechanics.

It was in trying to understand the structure of the periodic table that Danish physicist Niels Bohr (another Nobel Prize winner) came up with his model of the atom, where electrons were allowed to orbit in 'shells' around the centre, which were stable when filled with a certain number of electrons (sometimes two, sometimes eight, depending on the position of the shell). This model was uncovered by chemistry experiments, rather than through theoretical means, as it was found that elements with an even number of electrons were more stable than those with an odd number.

Explaining this theoretically was what Austrian physicist Wolfgang Pauli set out to do: what was so special about two or eight electrons in the same orbit? Pauli was one of the pioneers of quantum physics. His dad was a chemist, his sister a writer and actress, and his godfather was the one and only Ernst Mach (as in supersonic speeds measured in units of mach). Surrounded by such overachievers, I can only imagine the pressure Pauli put on himself to succeed. But succeed he did; if you have no idea who I'm talking about, let me say this: Einstein nominated Pauli for a Nobel Prize, which he won.[41]

In 1925, Pauli delved into how quantum mechanics describes electrons and realised that the elements of the periodic table could all be explained with just four quantum properties of electrons to describe their 'state': energy, angular momentum, magnetic moment and spin. The rule is that no two electrons around an atom could have the same values for those four properties. This is what's known as the Pauli exclusion principle; essentially it says that no two electrons can be in the same quantum state, i.e. have the same values for their four quantum properties. This is why each element in the periodic table is unique: because the electrons in their atoms have specific configurations defined by quantum mechanics that are replicated by no other element. Pauli figured out this

---

41 He also has the 'Pauli effect' named after him, whereby technical equipment seems to break around certain people. There were many anecdotes of fellow physicists complaining that their demonstrations would always fail when Pauli was around. The German-American physicist Otto Stern reportedly went as far as banning his friend Pauli from his lab. Perhaps I should've cited the Pauli effect when explaining to my chemistry teacher at Bolton School Girls' Division why boiling tubes and beakers always ended up smashed after I used them in a lesson.

one simple rule that explained the structure of all atoms and why some were more stable than others. This is why physicists like to joke that the entirety of chemistry can be explained in one page of quantum mechanics, to the intense frustration of chemists everywhere.

What the Pauli exclusion principle means for astrophysics is that if you squash a load of electrons under gravity, they'll resist being squished as there is no lower quantum state for them to go to; other electrons have already filled those states. This resistance is known as electron degeneracy pressure, and by 1926 the British astronomer Ralph Fowler[42] applied this new quantum mechanics discovery to the decades-old problem of the densities of white dwarf stars. He realised that the huge densities of white dwarfs, at around a billion kg/m³ (for context, water has a density of 1000 kg/m³), could be explained if gravity had crushed down the matter in stars so much that the electrons started to push back against gravity. Like many problems in science, though, solving this one led to a whole host of other questions. Including if there was a point when the electron degeneracy pressure was no longer able to resist that crush of gravity inwards. More simply, what was the maximum mass of a white dwarf?

---

42 Remember, the one who married Eileen Rutherford? Rather than William Fowler of B²FH paper fame (see page 54). Ralph Fowler was one of many twentieth-century physicists who were caught up in the First World War; he served in the Royal Marine Artillery of the British Army. His shoulder was wounded during the Gallipoli campaign, after which his physics talents were put to good use studying the aerodynamics of spinning anti-aircraft bombs.

It was Indian astrophysicist Subrahmanyan Chandrasekhar who cracked this one. Another overachiever, Chandrasekhar wrote his first scientific research paper aged nineteen during his undergraduate degree at the University of Madras. He sent this paper to Ralph Fowler at Trinity College, Cambridge, who promptly invited him to come and do a PhD at the university (Chandrasekhar was thankfully awarded a scholarship by the Indian government to pursue his graduate studies). Fowler had already attempted to determine what the limit to a white dwarf's mass might be, but Chandrasekhar, on his travels from India to the UK, realised Fowler's work needed some corrections using Einstein's theory of special relativity; the electrons had so much energy that their masses started increasing. I can only imagine Fowler's reaction when his new PhD student arrived with the news that he'd already cracked the problem Fowler had been working on for years. Over the course of his PhD, Chandrasekhar diligently revised his theory to give us what we now know as the Chandrasekhar limit for white dwarfs: 1.44 times the mass of the Sun.[43]

However, the idea of the Chandrasekhar limit was not well received by the astronomy community at the time, due to what it implied. Arthur Eddington was particularly vocal about it (the Big Name in Physics who had reasoned that stars could only be powered by nuclear fusion before there was any evidence for it). Eddington was also at Cambridge when

---

43 In his first paper on the limit in 1931, Chandrasekhar incorrectly concluded the limit was 0.910 times the mass of the Sun. A nice reminder of 'if at first you don't succeed, try, try again.'

Chandrasekhar completed his PhD, before being elected a new fellow of Trinity College in 1933 at the age of just twenty-three. Eddington was fifty-one and an eminent professor with international prestige, who used that influence to convince his colleagues that the idea of a limit to a white dwarf's mass was absurd. He went as far as presenting immediately after Chandrasekhar at a meeting of the Royal Astronomical Society in 1935, claiming that Chandrasekhar's theory was incomplete since it used two separate branches of physics: relativity and quantum mechanics (an argument that Pauli himself dismissed).[44] Eddington claimed that if we had a quantum relativity theory then the maths would come out to support *his* theory that white dwarfs were the last stage in the evolution of stars. He famously stated at that meeting: 'I think there should be a law of nature to prevent a star from behaving in this absurd way!'

Eddington, being the more senior academic, was taken more seriously than Chandrasekhar, who then had to fight for a good two decades before his theory became accepted, with both Chandrasekhar and Fowler eventually winning the Nobel Prize in 1983 (I do love a happy ending). Apart from his own ideas on stellar collapse being proven wrong, what

---

44 Eddington went about this in an academically brutal way; the minutes of this particular meeting of the Royal Astronomical Society read like a soap opera. Many have questioned whether Eddington's behaviour was motivated by race, but there are similar stories of scientific clashes with other junior researchers like Edward Arthur Milne (who studied how temperature changes in the atmospheres of stars) and James Jeans (who was one of the founders of modern cosmology).

was Eddington so worried about? Eddington thought it was absurd that there was a limit beyond which matter in a white dwarf star could not resist the crush of gravity, because what in the Universe could possibly happen next?

Eddington's fears were allayed for a few years by the discovery of the neutron in 1932 by James Chadwick (again in Cambridge at the Cavendish Laboratory[45]), completing the trifecta of the basic building blocks of all matter: electrons, protons and neutrons. This discovery led Walter Baade and Fritz Zwicky (two giants of astronomy from Germany and Switzerland respectively) to propose the existence of stars made entirely of neutrons just one year later in 1933. Here was an explanation for the next stage in the evolution of white dwarf stars after they get too massive and collapse under gravity.

Baade and Zwicky were working on a different problem, though; explaining what's left behind in a supernova. White dwarfs are formed when stars fizzle out, but explosive supernovae needed another explanation. They claimed this explanation was neutron stars. These neutron stars would be supported by neutron degeneracy pressure – similar to the pressure from electrons holding up white dwarf stars, neutron stars were held up by the inability of two neutrons to occupy the same quantum state, again according to the Pauli exclusion principle.

But just like with white dwarfs, the inevitable question of whether there was a limit to the mass of a neutron star reared its head. A mass so great that neutron degeneracy pressure

---

45 Seriously, what did they *not* do there?!

could not resist the crush of gravity inwards (the concept that Eddington found so absurd). This was tackled at the University of California, Berkeley, by American physicist Robert Oppenheimer[46] and his PhD student at the time, Russian-Canadian physicist George Volkoff, using previous work by Richard Tolman. In 1939 they derived the first estimate for what is now known as the Tolman–Oppenheimer–Volkoff limit for the maximum mass of a neutron star (the sibling of the Chandrasekhar limit), beyond which they claimed there was no known law of physics that would prevent the collapse of a star down to an infinitesimally small point with infinite density.

Eddington and many, many others were still not convinced, believing the notion of a gravitationally completely collapsed star (i.e. a black hole) to be completely unphysical. First, because no neutron stars had yet been discovered, and second, because the idea of a black hole, or of mass condensed into an infinitely small point, were just theoretical curiosities for the mathematically minded to ponder over. We can speculate whether, if Eddington had instead embraced Chandrasekhar's ideas and the application of the Pauli exclusion principle, he may have had a different role in this chapter, perhaps becoming the first physicist to predict the existence of a black hole, in the same way he predicted that nuclear fusion must be powering the Sun. Instead, the astronomical community came to begrudgingly accept the existence of black holes in

---

46 Of Manhattan Project infamy in the Second World War; Oppenheimer was one of the few who observed the Trinity Test in 1945 when the first atomic bomb was detonated. Again, knowledge of nuclear physics and neutrons has many applications.

Eddington's absence after a number of discoveries and observations later in the twentieth century.

First, in 1967 a PhD student at Mullard Radio Astronomy Observatory at the University of Cambridge, Jocelyn Bell,[47] working with Martin Hewish, discovered an unexplained radio signal which pulsed every 1⅓ seconds.[48] The following year then saw the discovery of the same repeating radio pulses from the centre of our old friend the Crab Nebula, the remnants of the AD 1054 supernova recorded by Chinese astronomers. By 1970, fifty pulsating radio sources had been found, and the explanation that was most favoured was of

47 In 2018, Dame Jocelyn Bell Burnell was awarded the Special Breakthrough Prize in Fundamental Physics, worth $3 million. She donated all the money to a grant 'to fund women, under-represented ethnic minority and refugee students to become physics researchers', which I think best summarises the wonderful person Jocelyn is. When I arrived in Oxford on my first day as a PhD student I was told that if I had any worries or concerns that I couldn't talk to my supervisor or college about, Jocelyn was the astrophysics department's 'ombudsman', and her door was always open for a friendly chat. You can just tell that she genuinely cares.

48 Martin Hewish went on to win the Nobel Prize in 1974 for his role in this discovery, along with Martin Ryle for their work pioneering radio astronomy. There is a rather large controversy around the fact that Bell Burnell was not included in the prize, especially given the fact the prize can be shared between a maximum of three people, but was only split between Hewish and Ryle. However, Bell Burnell herself said in 1977: 'I believe it would demean Nobel Prizes if they were awarded to research students, except in very exceptional cases, and I do not believe this is one of them.' I disagree with Jocelyn here; hindsight and science history have shown us that her discovery truly was one of those exceptional cases.

spinning neutron stars. These 'pulsars'[49] were the missing piece of the puzzle of understanding how stars end their life. Unfortunately, Eddington didn't live to see the discovery of neutron stars (having died from cancer at the age of sixty-one in 1944[50]), but the rest of the astronomical community realised what this discovery meant: if neutron stars were real objects, then perhaps black holes were not as *unnatural* as first thought. Coinciding with Bell Burnell and Hewish's pulsar discovery, in 1969 British physicists Roger Penrose and Stephen Hawking published a very mathematics-heavy paper showing how this gravitational collapse down to an infinitely dense, infinitesimally small point was actually inevitable in nature.

This all culminated in the release of a paper in 1972 by Australian astronomer Louise Webster and British astronomer Paul Murdin, who worked together at the Royal Observatory Greenwich to observe the mysterious X-ray and radio source Cygnus X-1. They observed a normal star that was found in the same part of the sky as Cygnus X-1 and noticed that the light from the star was Doppler-shifted. We all encounter Doppler shift in our day-to-day lives on a regular basis. As ambulance sirens race towards and away from us we hear the

---

49 According to Bell Burnell, 'pulsars' was an invention of *The Daily Telegraph*'s science reporter Anthony Michaelis. He suggested during an interview that since they'd been trying to study quasars (shortened from quasi-stellar objects) at the time, why not shorten 'pulsating radio objects' to 'pulsars'? And the name stuck.

50 Henry Russell (of the Hertzsprung–Russell diagram) wrote his obituary for the *Astrophysical Journal*.

change in pitch of the soundwaves as they are squashed to a smaller wavelength (or higher frequency) moving towards us, and stretched out to larger wavelength moving away from us. You can hear this at racetracks as cars zoom past and on motorway bridges as cars thunder underneath. This happens because sound is a wave. Just like sound, light is also a wave, and so the same process of squishing and stretching can also happen to light. As the wavelength is stretched to longer wavelengths the light becomes redder (redshift), and as it's squashed to shorter wavelengths the light becomes bluer (blueshift).

The star that Webster and Murdin observed was redshifted and blueshifted periodically every 5.6 days. This is caused when a star has a companion, so that the two orbit a centre of mass in empty space somewhere between the two. From how much the light is shifted, you can tell how fast the star is orbiting its companion and hence how heavy the companion to the star is, whether planet-sized (this is how we find a lot of Jupiter-sized planets) or *much* heavier. They calculated that this star's companion (which couldn't be seen) was greater than the theoretical Tolman–Oppenheimer–Volkoff limit, and that's when alarm bells started ringing. The paper they published with these measurements ends with the wonderful line: 'it is inevitable that we should also speculate that it might be a black hole.'

And so, by the 1970s, the trifecta of the graveyard of stars was complete: white dwarf, neutron star, black hole. Once a massive star, around ten times the mass of the Sun or larger, runs out of fuel, there's no process to stop the inevitable pull of gravity inwards on its core during the supernova and the only eventuality is that the core is crushed down into a black hole: a dark star. Today, we even think that some incredibly massive stars have directly collapsed into black holes and skipped supernova entirely, just – poof! – there one day and gone the next.

With the Chandrasekhar limit we also know that, in very special cases where they're given a supply of extra mass to grow, white dwarfs could one day collapse into neutron stars if enough mass is somehow added to them (eventually the electrons are forced to combine with the protons to make the neutrons that make up neutron stars). Similarly, with the Tolman–Oppenheimer–Volkoff limit we know that neutron stars could also one day become black holes if given enough mass to grow. This can actually occur if either the white dwarf or neutron star are in binary systems with other stars, which they can steal enough mass from to reach those limits. It's for this reason that I like to think of neutron stars as the prior evolutionary stage to a black hole: a Pikachu to a black hole's Raichu.

So, if we are willing to wait long enough, and there's a bumper supply of extra matter hanging around the Solar System neighbourhood, then theoretically, the Sun could one day become a white dwarf, grow into a neutron star and eventually a black hole. But that's true for just about any patch of gas in the Universe if you're patient enough to follow the recipe:

- Preheat oven to nuclear fusion temperature.
- Keep adding sprinkles of billions of kilograms worth of matter.
- Bake until crushed.

6

Funny, it's spelled
just like 'escape'

Some of the best days of my life have been spent travelling the world. I vividly remember a trip to the Grand Canyon when I was a teenager; the colours, the heat, the sheer scale of the thing took my breath away. I crept ever closer to the cliff edge, just to catch a better glimpse of what it had to offer. The closer I got to the edge, the more I could see of the canyon itself; the strange rock formations and the water meandering through its base. Of course, being a teenager at the time, I could not be trusted; my parents reminded me every five minutes not to get too close to the edge, and being a teenager I terrorised them and did it anyway. But, as parents often are, they were right to warn me (not just because I am possibly the clumsiest person they've ever known), because if I'd taken just one step too close to that edge, I would have fallen a long way down.

Let's presume I'd have survived such a fall to the bottom of the Grand Canyon – I would've then been stranded at the bottom of the valley without enough energy to claw my way back up the cliff face. Now, I realise I've spent the last few chapters convincing you that black holes aren't holes but mountains instead, but you can picture what's known as the 'event horizon' around a black hole as the edge of the Grand

Canyon – it's the point at which you've gone too far and neither you nor anything else in the entire Universe has enough energy to claw its way back out.

As you get closer to a black hole, the escape velocity needed increases until it reaches the speed of light. This point is what we call the event horizon, and it only exists because of that ultimate speed limit in the Universe – the speed of light. The event horizon is often described as the 'point of no return' – but it's not a *point* at all. The event horizon is a three-dimensional sphere around whatever lies inside, and it's what we describe as the 'size' of a black hole, known as the Schwarzschild radius.

Karl Schwarzschild was a German physicist and astronomer, who at the outbreak of the First World War was director of the Astrophysical Observatory at Potsdam.[51] Despite being exempt from mandatory service in the German army due to his age (he was pushing forty-one), he volunteered and served on both fronts. For Schwarzschild, though, war did not put a stop to his science, because in the middle of the First World War, in 1915, Einstein announced his theory of general relativity to the world, including the equations that described how space and time were affected when matter was present. These equations are notoriously tricky beasts to solve,[52] and even Einstein himself didn't think they had exact solutions, having made many approximations himself to get answers (for

---

51 That's a really big deal.

52 To the dismay of physics students around the world, there is no equivalent to the 'quadratic formula' – $x = \frac{-b \pm \sqrt{b^2 - 4ac}}{2a}$ – for Einstein's field equations. If only.

example to explain the orbit of Mercury). But that didn't put off German army artillery lieutenant Karl Schwarzschild, who is a figure from history who could definitely be described as 'non-stop'.[53]

During his time serving on the Eastern Front (and despite suffering from a rare, painful autoimmune disease), Schwarzschild wrote *three* scientific papers in his 'down-time', two of which were on general relativity.[54] He worked out an exact solution to Einstein's field equations for the strength of gravity around a spherical, non-rotating object using the simple trick of employing a different coordinate system (instead of normal x, y, z coordinates, he used polar coordinates of radius and angle, like you would use to describe a position on the Earth in terms of latitude and longitude). After he figured out this solution to the equations, he wrote Einstein a letter on 22 December 1915, while he was still on the Eastern Front. There's a fantastic line in it, which in its original German reads: '*Wie Sie sehen, meint es der Krieg freundlich mit mir, indem er mir trotz heftigen Geschützfeuers in der durchaus terrestrischer Entfernung diesen Spaziergang in dem von Ihrem Ideenlande erlaubte.*' He is thanking Einstein here, saying that the war has treated him kindly enough despite all the heavy gunfire, as it has given him the opportunity to take a walk through Einstein's ideas about gravity published in his theory of general relativity. This line is especially poignant

---

53 Another one for all my *Hamilton* fans out there.
54 'Why do you write like you're running out of time?' I can't stop. Love you Lin-Manuel.

with the knowledge that Schwarzschild died just five months later in May 1916, at the age of just forty-two.

Schwarzschild wasn't trying to solve these equations for a black hole; instead his solutions describe any sphere of mass, whether a star or a diffuse nebula of gas scattered across a huge region of space. But there was something about his solution that had people worried for decades afterwards, because in this solution there were two points at which the strength of gravity became infinite. Because Schwarzschild used polar coordinates, the equation he got for the strength of gravity depended on the distance away from a certain central point. So, the solution is the same for all points at this distance, i.e. a sphere defined by a certain radius. When that radius was equal to zero, the strength of gravity became infinite. But it also occurred at a larger distance too; a distance that depended on the mass.

The places where this happened were known as 'singularities'. This is a fancy mathematics word that means 'we can't tell you what happens here'. It's a point that is undefinable, that's usually *reviled* by mathematicians because to work out the strength of gravity at a radius of zero, you have to *takes deep breath* divide by zero (I think I just involuntarily shuddered). Dividing by zero is mathematically impossible, but us physicists don't dwell on it for too long. If you take ever smaller and smaller numbers and divide by them, the answer you get grows and grows (like with an object's momentum as it travels closer to the speed of light). So us physicists are very happy to divide by zero and say that we get infinity, something mathematicians will debate the philosophy of endlessly. Now, these singularities weren't an issue for most objects, like stars, as the larger radius, where the other singularity appeared, is very small and normal stars are very large. This radius is now known as the Schwarzschild radius, but it wasn't until the 1960s that it would be recognised for what it truly was: an *event horizon*.

It was Austrian-born physicist Wolfgang Rindler who we have to thank for the term 'event horizon'. At the age of just fourteen, Rindler was evacuated from Austria to England in the Kindertransport rescue of Jewish children before the outbreak of the Second World War. He finished school and went to university in the UK, before being offered a job at Cornell University in New York state in 1956. Once at Cornell, Rindler managed to publish the results from his PhD research at the University of London, and the world was introduced to the idea of an event horizon. He defined a 'horizon' as 'a frontier between things observable and things unobservable', in the same way you can't see anything beyond the Earth's horizon when looking into the distance. An event horizon therefore divides events into those that can be seen, and those that can't. Or, to put it in Rindler's much more poetic words: 'those [events] that are forever outside [our] possible powers of observation'.

The Schwarzschild radius is the event horizon of a black hole; it marks the region where we no longer get any information out of the black hole because light can no longer reach us. It is the region where the escape velocity becomes greater than the speed of light. It's not a true singularity, because if you use a different coordinate system you can still define the strength of gravity at that point, even if you can't get any real information from beyond it. But the Schwarzschild radius still represents something physical about black holes. Schwarzschild's solution to Einstein's general relativity equations essentially tells us the size of the event horizon, or the size of the black hole itself. The size is only dependent on the mass of the black hole (and the speed of light and the overall strength of gravity, but those are constant values that don't change, as far as we know). Essentially, the bigger the black hole, the bigger the event horizon.

I remember first learning the derivation of Schwarzschild's solution while I was studying for my undergraduate degree in physics at Durham University. Obviously, once armed with the equation to calculate the size of a black hole, the first thing I did was work out how large of a black hole I personally would be. Having consumed a fair bit more cheese since my university days I've had to re-do the calculation for this book, but in case you were wondering, if we had the ability to squish an average human at around 62 kilograms down into a black hole, they would have an event horizon with a radius of about 0.09 yoctometres (0.00000000000000000000 000009 m; that's twenty-five zeros after the decimal place!). That's smaller than an atom. Smaller than a proton that makes up the nuclei of atoms. Smaller even than the quarks that make up protons.

Admittedly, it's a number that is a little bit small for our brains to comprehend, so let's try something bigger: the entire Earth, for example. If you could take the Earth and turn it into a black hole it would have a radius of just 0.9 cm, smaller than your fingernail. Whereas if you could turn the Sun into a black hole it would have a radius of 2.9 km. The actual radius of the Sun is 696,342 km, much larger than its Schwarzschild radius. But no matter the size, whether 0.09 yoctometres, 0.9 cm or 2.9 km, the black holes we could make out of the Earth and Sun would behave in exactly the same way, with escape velocities higher than that finite speed limit of the Universe: the speed of light.

But what about the other singularity in Schwarzschild's solution? The one that appears at $r = 0$. The Schwarzschild radius isn't a real singularity, it's what's known as a 'coordinate singularity' (it only exists because of whatever system of coordinates you've solved your problem in), but the one at $r = 0$ is a genuine physical singularity known as a 'gravitational singularity'. It is completely undefinable and unknowable. The curvature of space at that point, and therefore the strength of gravity, cannot be defined. In fact, the point itself is not even considered to be a part of normal 'spacetime' anymore; you can't define where (or even when!) the point is.

Again, this isn't a big deal for objects that are much larger than the Schwarzschild radius, like for a star whose mass is nice and evenly distributed. We don't need to know the value at $r = 0$ and we can say that the strength of gravity is nicely described by Schwarzschild's solution to Einstein's equation, as long as r is greater than 0. It is a big deal, though, when we think about the end of a star's life, when there's so much mass in the core that nothing is able to resist the crush of gravity. Not electron degeneracy pressure, nor neutron degeneracy pressure. The star keeps collapsing, getting ever smaller, until it becomes smaller than the Schwarzschild radius. What happens to it then? We don't know, because the star's collapse is now an event that is occurring beyond the event horizon: *forever outside our powers of observation.*

There is no process or form of matter that we know of in all of physics that can resist gravity to stop the collapse of the star. As far as we know, it keeps collapsing down to an ever smaller size until all the mass is contained in an infinitely dense, infinitesimally small undefinable point at $r = 0$: the singularity. At least this is the mathematical description. The event horizon shrouds the true nature of what's 'inside' the black hole from our view, due to the nature of light itself: what do these dark stars truly look like?

Light is how we observe the Universe around us; recording the brightness of stars or the positions of planets reflecting light from the Sun. We send information encoded on radio waves of light through the air which get decoded into sound at the other end. We do the same with infrared light through fibre-optic cables so we can access the internet. We communicate and receive information with light. This means black holes are not only prisons for light, but prisons for information and data. Under the laws of physics, as we understand them right now, we might be able to run the maths beyond the event horizon as much as we like, but we can never test those predictions because we can never receive any information from beyond the event horizon of a black hole.

No data equals very sad scientists. Imagine the feeling if you got closer to the cliff edge of the Grand Canyon and yet you still couldn't see into the spectacular canyon itself. It's utterly infuriating. But it's something us astronomers have had to resign ourselves to. However, unlike the Grand Canyon, which has a very obvious and clear cliff edge that has likely been making parents nervous for thousands of years, the event horizon is not obvious at all. There's no cliff edge around a black hole. No line drawn in the sand. No Schwarzschild dressed as a referee with a spray can drawing a line on the pitch. An event horizon is completely and utterly invisible, you wouldn't even know it was there unless you were paying close attention . . . adventurous space travellers beware!

7

Why black holes are not 'black'

I t never ceases to amaze me that we can actually see the
stars in the night sky. That might sound like a rather silly
thing for an astronomer to say, but really sit and *think* for a
moment at just how far starlight has had to travel before finally
reaching our eyes. Next time you catch a glimpse of the night
sky, see if you can find the three stars in Orion's Belt. The
closest star in Orion's belt is 11 quadrillion kilometres away,
or 1,200 *light years*. That means it took that light 1,200 years
to travel from that star to our eyes.[55] Not only are we seeing
the star as it was 1,200 years ago, but somehow, a tiny part of
the light which was sent out in all directions across the Universe
has managed to make it to our eyes across such a vast distance.

Think of how lights like torches and car headlights get so
much fainter when we get further away from them. So now,
just stop to picture how bright those stars truly have to be
for us to be able to spot them with a quick glance towards
the sky from our bedroom windows, despite being quadrillions

---

55 The other two stars in Orion's Belt are 1,260 and 2,000 light years
away. A reminder that although the stars in constellations appear close
together when projected onto the two-dimensional night sky (appearing
as if they are points on the inside of a sphere), in reality, in three dimensions, they are literally light years apart.

of kilometres away and competing with the glare of the street lamp across the road. This is why I catch my breath every time I look at the sky. I get lost in the knowledge of how easy it is for us to merely look up and see even tiny pinpricks of light that have been on the most epic of journeys.

Every star you can see in the night sky is in our local neighbourhood of the galaxy. The light from the stars further away in the Milky Way, over on the other side of the galaxy, combine together in one big faint fuzzy glow that looks like someone has spilled milk across the sky (hence how our galaxy originally got its name; the word galaxy even comes from the Greek *galakt*, meaning milk). Those of you who have seen a truly dark sky, away from the light pollution of cities and towns, won't have been able to miss the arch of the Milky Way (it's a flat spiral shape, with all the stars orbiting in one plane like the planets in the Solar System, so it appears as a strip across the sky), and those of you who have only ever seen the night sky from a city will probably not know what I'm talking about. Fainter still in the night sky is the galaxy Andromeda, which is made up of over a trillion stars; it's visible from the northern hemisphere as a small fuzzy blob, but in reality it extends about the width of six full Moons across the sky. It's just so far away that the light from those trillion stars is so faint we can barely just detect it with our eyes.

The view is very different when you break out a telescope, though – when Galileo trained his telescope on the fuzzy glow of the Milky Way back in the 1600s, he was shocked to see it resolve into lots of individual stars. Telescopes have allowed us to see further and fainter things in greater detail than we could ever see with our own eyes. And not just tele-scopes that see in visible light, like our eyes do, but those that detect radio waves (like those used by Bell Burnell and Hewish to discover pulsars) and also incredibly energetic X-rays.

As we saw earlier, X-rays and radio waves are all forms of light; just light with different wavelengths across the entire spectrum. The rainbow doesn't end at red and violet, it's just that our eyes can't detect the light beyond those colours. It was Scottish physicist James Clerk Maxwell who made that jump in understanding of what's actually 'over the rainbow' in 1867. Maxwell's equations, as they're now known, are the foundation of every single physics university course the world over. They explain what light is; a wave made up of an elec-tric and a magnetic part (an electromagnetic wave), and how these waves travel. Maxwell concluded that visible light was an electromagnetic wave with a very short wavelength, and predicted the existence of other electromagnetic waves with both longer and shorter wavelengths with different properties.

Maxwell's equations were just that though: equations. Mathematics only. No one had yet proved that light was actually an electromagnetic wave, or observed those with longer or shorter wavelengths that Maxwell had predicted. But just twenty years later, in 1887, a German physicist named Henrich Hertz invented a device that would generate what we now know as radio waves; electromagnetic waves with a much longer wavelength than visible light. Over the next few years he would go on to prove that they behaved just as Maxwell had predicted, and crucially behaved the same way that visible light did. They reflected, refracted (changing direction when passing from e.g. air to glass, as Fraunhofer found to his intense frustration) and diffracted (spread out around an obstacle or opening, like ocean waves in a cove).

Hertz's discovery wasn't just the first ever recorded generation of radio waves, it was the very first proof of Maxwell's equations and ideas about what light actually is. It opened the door for the discovery of even more types of electromagnetic radiation; and in particular for the 'accidental' discovery of X-rays in 1895 by another German physicist, Wilhelm Röntgen. Röntgen was working at the University of Würzburg, playing around with some of Thomson's cathode ray tubes. As Thomson later discovered, cathode rays are essentially a stream of electrons flowing from a negatively to a positively charged rod of metal. The electrons are accelerated to huge speeds by the voltage difference between the two rods.

Electrons are tiny particles, invisible to the naked eye, so we can't see the actual cathode ray itself, but what people noticed at the end of the nineteenth century was that if the electrons hit the inside of the glass tube, the glass would glow. The atoms in the glass were absorbing some of the electron's energy and emitting it as light – this is fluorescence.

Röntgen was trying to establish if it was possible to get the cathode ray out of the tube through a little opening in the glass (the opening was made from aluminium, to block light but conduct electrons). He figured if he covered the entire thing in thick black paper to shield any of the fluorescent glow from inside of the glass, he could then observe if he saw any fluorescence outside of the opening in the glass. To check if his paper cover was completely light-tight, he covered the aluminium opening with his black paper and then turned off all the lights in his lab. He didn't see any fluorescent glow coming from his paper-sheathed creation and so, satisfied, he went to turn the light back on. It was then, in the dark of the lab, that he spotted something shimmering on a bench a few metres away from the tube. Far further than anyone expected the cathode ray to be able to travel through air; quite famously, electrons need a good conductor to travel through, like copper, hence why all our houses are full of copper wire (or even copper-sheathed aluminium) to deliver us precious electricity.

Röntgen, not believing what his eyes were seeing, repeated the experiment a few times, running a voltage through the paper-sheathed glass tube repeatedly before being convinced this fluorescence was real. He determined that the fluorescence must be caused by a brand new type of radiation. Since these rays were unknown to him, he used the classic mathematical symbol for an unknown property: 'x', and dubbed them 'X-rays'. That term has stuck, at least in English in anyway – in many European languages X-rays are actually known as Röntgen rays.

He then dove into understanding as much as possible about these new 'X-rays'. What materials could they travel through? How much fluorescence did they cause? How were they generated? He recorded all of this with photographic plates; in the early days of photography images were created by exposing metal plates coated in silver-based salts that were sensitive to light. Where light hit the plate, the substance would turn dark (we know this today as a negative image). His biggest breakthrough was when he moved a piece of lead in front of the opening of the cathode ray tube and noticed it blocked the X-rays, along with his own hand. After seeing a ghostly image of his own hand on the photographic plate, he started to conduct his experiments in secret, fearing his scientific reputation was on the line. However, other scientists had already noticed that photographic plates became exposed if left too close to a cathode ray tube, with American physicist Arthur Goodspeed noticing that a photographic plate with two coins left on it developed to show two dark circles.

So Röntgen, despite his doubts, decided to continue investigating which substances blocked these 'X-rays' and which didn't. It fell to his wife, Anna Bertha Ludwig, to act as guinea pig in his experiment, and he managed to capture the very first recognisable medical X-ray of the bones in her hand. Her bones and the ring on her finger blocked more X-rays than the muscle and skin surrounding them, and so appeared darker on the image. The image looks so familiar and recognisable to us in the twenty-first century (an X-ray is barely anything to blink at when they appear in the background of an episode of *Grey's Anatomy*), but on seeing the image of her skeletal fingers, the first of its kind, Anna Bertha is reported to have said, 'I have seen my death!'

By December 1895, Röntgen had published his work and the discovery of this new kind of radiation took both the public and the scientific world by storm. Practically every physicist had a cathode ray tube in their lab at that time, which meant they could drop everything to recreate Röntgen's experiment and further study these mysterious new rays themselves. But it was Röntgen himself that recognised how useful they would

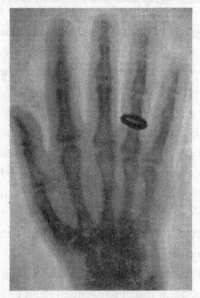

The first ever X-ray published by Wilhelm Röntgen in 1896, of the hand of his wife, Anna Bertha Ludwig. Darker areas are where bone and jewellery block more X-rays. Lighter areas are where fewer X-rays are blocked.

be in medicine, writing letters about his discovery to every doctor he knew. Within a year, the medical community across the world was using X-rays to locate bullet fragments, see bone fractures, locate comical swallowed objects and more (although with a bit more of a devil-may-care attitude than nowadays, as they didn't know the dangers that continuous exposure to high doses of X-rays can pose[56]).

---

56 Even up until the late 1940s, shoe shops would offer free X-rays so customers could see the bones in their feet.

It wasn't until 1912 that Max von Laue (another German physicist), together with his students doing the grunt work, would figure out what Röntgen's X-rays were: an electromagnetic wave. They were light, but with a much shorter wavelength than visible light; generated when the electrons in the cathode ray collided with the aluminium covering the opening in the glass tube, and then carried on unhindered through the heavy paper covering it. Röntgen never patented his discovery on ethical grounds, believing something so beneficial to medicine should be free to all. He eventually won the very first Nobel Prize in Physics in 1901 for the discovery, donating the 50,000 Swedish krona prize to further research at the University of Würzburg.

Röntgen's discovery might have rocked the physics and medical worlds, but it didn't have much of an effect on astronomers for another fifty years. Max von Laue's discovery that Röntgen's rays were a type of light might have planted the idea of observing the sky with X-rays in astronomers' minds, but it was far from feasible. Thankfully for life on this planet, Earth's atmosphere blocks the majority of harmful X-rays from outer space from reaching us down on the surface (unlike visible light and some radio waves, which make it through no problem). Great news for us: bad news for budding X-ray astronomers in the early twentieth century.

The atmosphere makes the process of detecting X-rays from objects in space harder than detecting optical light, UV light or radio light. You can't just cobble together the parts for a telescope on the university's spare patch of land. Instead, you have to launch your telescope, along with an X-ray detector, up above the Earth's atmosphere. Sounds fairly easy now to you and me, who are used to even private space companies launching satellites, spacecraft or perhaps the odd electric car into space nearly every single day. But back in the early twentieth century, the idea of X-ray astronomy was considered by most astronomers as just too much of a faff.

Not so for Riccardo Giacconi, though, who having seen the leaps and bounds in knowledge made in the physics world thanks to X-rays, made X-ray astronomy his mission. Giacconi was an Italian-American astronomer who, after completing his PhD at the University of Milan in 1954, jumped ship to the USA on a Fulbright Scholarship.[57] Giacconi had been captivated by earlier efforts to detect X-rays at ever-higher altitudes using balloons. But the time of the balloon was over; the time of the rocket-based X-ray observatory had come.

In a technique that would be used until the early 1970s, rockets with X-ray detectors attached would make short flights into the upper reaches of the Earth's atmosphere and back down again, recording any detections on the way. Giacconi's experiments using this technique revealed that there were X-rays peppering the night sky, appearing to come from areas

---

57 Today, the Fulbright Program is the largest international cultural exchange in the USA, operating in over 155 countries worldwide with over 294,000 scholarships awarded in the past sixty years to students wishing to study or teach abroad in a huge range of subjects. It has an enormous legacy, with eighty-eight alumni having received a Pulitzer Prize for journalism, sixty alumni winning Nobel Prizes (in physics, chemistry, medicine, literature and/or peace), thirty-eight having served as a head of state and one as Secretary General of the United Nations.

where there were no known visible objects. The question on everyone's lips was, *what could possibly be generating these X-rays?*

People were stumped, because there's not a lot of processes that have enough energy to produce X-rays. X-rays are an extremely short wavelength of light, so they're very energetic. They're only given off when something is extraordinarily hot (or moving very fast, like the electrons in a cathode ray tube). Not even the surface of the Sun, at 5,700°C, is hot enough to produce X-rays. The Sun's upper atmosphere (the corona), on the other hand, is millions of degrees and is plenty hot enough to give us X-rays (remember that the wavelength of light given off depends on the temperature).[58] The Sun's X-rays were discovered in 1949 by American X-ray astronomer Herbert Friedman during a rocket flight, and although the Sun is the brightest source of X-rays in our sky, that's only because it's so close. The Sun is not a very powerful source of X-rays, unlike the X-rays found strewn across the sky in Giacconi's experiments.

In 1962, Giacconi detected one of the strongest sources of X-ray light in the sky using the rocket-based method, coming from the direction of the constellation Scorpius.[59] Given the

[58] It's not very well understood why the atmosphere of the Sun is so much hotter than its surface. Hypotheses have ranged from the magnetic field of the Sun being to blame, to escaping radiation from tiny sunspots on the surface. I think it's a nice reminder that although we now know so many things about our wider Universe, there's still so much we don't know about even our own Sun.

[59] Remember, stars in constellations are actually light years apart, like in Orion's Belt. Something being in a constellation doesn't mean it's nearby to the other stars in the constellation, just that it's in the same

technology of the X-ray detector on board the rocket at the time, that was about as detailed in terms of location that you could get; definitely not from the Moon. It was announced to the world as the first X-ray detection from outside the Solar System. With further rocket flights the location was narrowed down to a star called V818 Scorpii, and the X-ray source, being the first discovered in the constellation of Scorpius, was dubbed Scorpius X-1. This led astronomers to debate whether other stars might be giving off more X-rays from the hot corona that surrounds them, as our Sun does, and this remained the most likely explanation for a few years.

That was until 1967, when the Soviet astronomer (born in what is now Ukraine) Iosif Shklovsky argued that explanation couldn't possibly be right. He said that stars just didn't have enough energy to produce that many high energy X-rays; they weren't hot enough. Shklovsky was a big name at the time, in both the scientific world and the public eye, having published a book in 1962 on intelligent life in the Universe, in his native Russian, that was then reissued in English in 1966 with Carl Sagan as a co-author.[60] Shklovsky was one of the five giants who pioneered the scientific search for intelligent life beyond Earth, along with Sagan, Italian physicist Giuseppe Cocconi and American astronomers Philip Morrison and Frank Drake (of Drake equation fame, and who was supervised by Cecilia Payne-Gaposchkin).

---

direction in that part of the sky from Earth's perspective. Constellations are just used by astronomers as handy marker points for navigating the sky, to give general directions of objects.

60 Both Shklovsky and Sagan shared Ukrainian-Jewish heritage.

By 1967, Shklovsky had had a thirty-year career focusing on high-energy astrophysics phenomena (from supernova remnants like the Crab Nebula to the Sun's corona emitting X-rays) while dabbling in the orbits of the moons of Mars and extra-terrestrial life. So, when he posed a new explanation for Scorpius X-1, people took note, even if at the time it seemed pure theoretical fancy. He concluded the only process that would have enough energy to produce the X-rays seen would be an accreting (accretion is a fancy physics word that means to grow gradually in mass) dense object, like a neutron star. Shklovsky published his paper in April of 1967, seven months before Jocelyn Bell Burnell would spot that bit of scruff in her data that marked the discovery of the first neutron star.

So how did Shklovsky make this leap in understanding? From the maths of how fluids (i.e. liquids and gases) behave, physicists had long known that gas moving incredibly quickly would heat up to equally incredible temperatures. Similarly, if that gas was all moving in one direction, perhaps orbiting something, they knew that it would form a disk shape, like how a ball of pizza dough can flatten out into a pizza shape when it's set spinning overhead (at least, by a very talented chef; mine always ends up on the floor). He suggested that the only scenario that could explain the energies of the X-rays detected was if Scorpius X-1 was a very dense object in orbit around, and also stealing matter from, the star V818 Scorpii. He argued that only a neutron star could be accreting this matter, accelerating it to huge speeds to form what's known as an accretion disk around it, and therefore heating it to extreme temperatures so that it gave off X-rays.

With the discovery of pulsars by Jocelyn Bell Burnell, and their eventual explanation as neutron stars, Shklovsky's hypothesis about Scorpius X-1 became all the more attractive, and the idea was eventually accepted by the scientific community at the start of the 1970s. The 1970s then saw a huge leap in the field of X-ray astronomy: space telescopes. Instead of launching rockets, scientists could launch a satellite into orbit with an X-ray detector on board. The very first was Uhuru[61] in December 1970, which surveyed the entire sky, marking the locations of X-ray sources and discovering many more sources that coincided with normal stars (including Cygnus X-1, the very first candidate black hole that we heard about in chapter 5) and with newly discovered radio sources like pulsars.

One source of note was Centaurus X-3 (the third X-ray source found in the constellation of Centaurus in the southern hemisphere sky), which was detected in X-rays first but was later found to be a pulsar giving off radio waves and orbiting a normal star called Krzemiński's star (after it's discoverer, Polish astronomer Wojciech Krzemiński). Centaurus X-3, along with many other X-ray sources like it, left no doubt in people's minds that these X-rays were powered by accretion,

---

61 Uhuru is the Swahili word for freedom. The satellite was named to honour Kenya, after being launched from near Mombasa. The closer you are to the equator for space launches the better; the equator is spinning faster than the Earth's poles, so you get an extra boost of energy. Anywhere with an Eastern coast is also preferable, due to the direction of the Earth's rotation; if anything goes wrong, your rocket crashes into the sea rather than onto land.

just as Shklovsky had suggested. In Centaurus X-3's case, the compact object is a neutron star – it's very clearly detected as a radio pulsar. But in some cases, like in Cygnus X-1, the X-ray energies were so huge, much larger than those seen coming from accreting neutron stars, that the only explanation was something much larger than the Tolman–Oppenheimer–Volkoff limit for the maximum mass of a neutron star. Cygnus X-1 could only be powered by an accreting black hole.

So, in the mid-1970s, Russian astrophysicists Nikolai Shakura, Rashid Sunyaev and Igor Novikov and American theoretical physicist Kip Thorne, first modelled how gas orbiting a black hole would heat up to anywhere from 10,000 to 10,000,000 kelvin depending on how massive the black hole (or other compact object) was. This accretion process essentially converts mass into energy (remember, because they are equivalent) in the form of light, which is also how nuclear fusion inside stars can be described. Accretion, though, is much more efficient than nuclear fusion. If 1 kilogram of hydrogen were to undergo nuclear fusion inside a star, only 0.007 per cent of that mass would be released as radiation. Whereas if 1 kilogram of hydrogen was accreted by a black hole, 10 per cent of that mass would be released as light as it spiralled towards the black hole in the accretion disk. That's the key thing here – the light is released from the accretion disk around the black hole, which is much further out from the event horizon, so we can still detect it.

It's these detections of X-rays that allow us to know that black holes, dead stars, are hiding out there among the stars of the Milky Way. Unlike what you might first think, black holes are terrible at saying hidden; they make the material around them light up like a Christmas tree. Because of accretion, black holes are not 'black' at all; they end up being the brightest objects in the entire Universe. So you're not reading a book about Robert H. Dicke's 'black holes', but one about blindingly bright mountains.

8

When 2 become 1

One of the wonderful things about the night sky is that it's available to everyone. At least, those who aren't plagued by bad weather. Anyone with clear skies can head outside, observe the sky with or without a telescope and run through the scientific method to try and explain the observations they have made. Advancements in technology have also made doing these observations far easier, from night-sky apps that tell you exactly what you're looking at, to telescopes and cameras that allow astrophotographers to capture images from their back gardens that would have been the dream of early twentieth-century astrophysicists. One thing technology has given us is the ability to 'see' without light. In a whole new way.

The majority of stars like our Sun aren't found alone. Our Sun is quite rare in that respect – more than 50 per cent of Sun-like stars are found orbiting another star. The two stars will orbit around a common *centre of mass*. If the two stars are exactly the same mass, the centre of mass will be perfectly in the middle and the two will orbit like two friends who spin around holding hands, perfectly equidistant from each other, following the same orbit. But if one star is heavier than the other, then the centre of mass will be offset. Picture the two

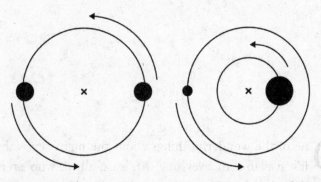

Diagrams of two stars of the same mass (left) and different masses (right) orbiting the centre of mass, marked by the crosses between them.

stars on a seesaw: if one star is heavier than the other, you'd have to move the pivot point from the middle to a point closer to the heavier star to get the seesaw to balance perfectly. That point is the centre of mass that they orbit around, meaning the heavier star traces a smaller orbit at a slower speed, and the smaller star a much longer one at a faster speed.

Two stars orbiting each other is known as a binary system, but you can throw in another star orbiting those two to give a tertiary system, or have two pairs orbiting the centre of mass between them in a quadruple system. The largest number of stars we've ever found in one star system (at least at time of writing) is a whopping seven. There are two seven-star systems that we know of; Nu Scorpii and AR Cassiopeiae.[62]

---

62 Unfortunately, they're just a *little* bit too faint to see with the naked eye, but with binoculars and a map of the constellations of Scorpius and Cassiopeia, you should be able to find both in clear, dark skies.

AR Cassiopeiae is a binary system, orbiting a binary system, orbiting a triple system. Nu Scorpii is slightly simpler, with a triple system orbiting a quadruple system.

The more massive the star, the more likely it is to be found in a multi-star system with companions. In the case of very small red dwarf stars (which are very low mass and faint but make up about 85 per cent of all the stars in the Milky Way[63]), only 25 per cent have a companion star, but that increases to more than 80 per cent for the most massive stars, which will collapse into black holes at the end of their lives. To form a massive star, you need lots of gas in one place, and so the majority of these form in big clusters of stars from a single giant gas cloud. So many stars in an astronomically speaking 'small' space increases the likelihood of massive stars ending up in multi-star systems.

The most massive stars also run out fuel quicker; as we learnt earlier, they live fast and die young. Because they're so massive, the crush of gravity inwards is huge, so they have to burn more fuel to counteract it and therefore run out a lot quicker. While the Sun will live for about 10 billion years (it's currently middle aged at around 4.5 billion years old), the most massive stars live for 100,000 years if they're lucky; burning the brightest for the shortest of astronomical times.

---

63 Before the realisation that red dwarf stars are far more common than first thought, astronomers were skewed by the fact that the more obvious, brighter, massive stars had a companion more often than not and believed that the majority of stars in the Milky Way were in multi-star systems. Instead, since red dwarfs make up the majority of stars, only a third of stars in the Milky Way are in multi-star systems.

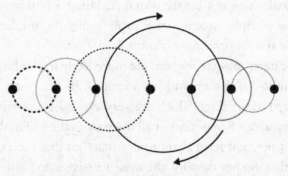

The set-up of seven-star system Nu Scorpii, with stars shown by the filled circles and their orbits shown by the rings. A tertiary and quadruple system orbit each other. The tertiary system is one star orbiting a binary system of two stars. The quadruple system is one star orbiting another star which orbits a binary system of two stars.

This means that more often than not you end up with a black hole (or neutron star or white dwarf) in orbit around a normal star which is still going to keep happily fusing hydrogen to make helium for many millions, if not billions, of years. This is what happened to our old friend Cygnus X-1 – the first ever candidate black hole from the previous chapters – and countless other systems.

We spot these binary systems containing black holes all across the Milky Way thanks to their X-ray light. But what if the second star in that system is also a massive star, which goes supernova and becomes a black hole? Then you'd end up with two black holes orbiting each other. The scale of the gravitational forces involved in this would be unfathomable. The stable orbit that the two stars would have been in during their lives would be completely disrupted by the two supernovae. A supernova throws the majority of the outer layers of the star, and therefore the majority of its mass, out into space, leaving only the core of the star to collapse into a black hole. So a new centre of mass would be needed to balance the system.

Since the second star going supernova throws off mass and results in a less massive black hole, that would mean the two black holes have to get closer together to find a new centre of mass. But there's no way two black holes will reach a stable orbit when they are that close together. What happens is that they eventually end up spiralling ever closer together over millions of years to an inevitable end in the most monumental collision you've ever seen. Or technically, *not* seen. Once both of the objects in the binary system are black holes, there's no material left to steal from a normal star to form an accretion disk which glows in X-rays. The whole system becomes completely invisible to us, at least until the very last moment.

Cast your mind all the way back to Chapter 3 and Einstein's theory of general relativity: mass curves spacetime. Massive objects, like black holes, have the most effect, curving spacetime to its extremes. As they spiral in towards each other, two black holes in a binary system accelerate along their orbits, constantly changing the curvature of spacetime around them. Changing the curvature of spacetime to the extreme, on a regular basis, takes a phenomenal amount of energy; energy that comes from the black holes themselves. It's like a shock to the system for space itself; it can't contain that much energy in such a small area so the energy disperses, rippling away across space like a shockwave.

If we go back to our analogy of massive objects as basket-balls on a trampoline, imagine bouncing two very heavy basketballs off the trampoline in a steady rhythm. The surface of the trampoline doesn't stay flat – it constantly bounces up and down as it absorbs energy from the bouncing basketballs, which ripples away across the surface of the trampoline. This is how we can imagine space, curved to the extreme and back by two black holes orbiting each other. The energy involved, that can't be contained in the area, is rippled away through space, like the ripples on the surface of a pond, as something known as a 'gravitational wave'. A wave through space itself, changing the curvature of space as it goes, fuelled by the energy injected by two orbiting black holes.

Einstein predicted the existence of gravitational waves way back in 1915 when he first published general relativity (although he didn't predict they'd be produced by black holes, but still by very dense, compact objects), but their existence would have to wait to be proven, at least indirectly, for another fifty-nine years. In 1974, two American astrophysicists, Joseph Taylor and Russell Hulse (both at the University of Massachusetts Amherst; Taylor was a professor and Hulse his PhD student), discovered the very first binary pulsar system, dubbed PSR B1913+16 (although it's now known as the Hulse–Taylor binary). This system consists of two neutron stars in orbit around each other, formed after the massive stars that preceded them had gone supernova.

Hulse and Taylor were using the Arecibo telescope at the time: a huge 305-metre radio dish in Puerto Rico, most famous outside astronomy research circles for starring in the 1995

James Bond film *GoldenEye* and the 1997 film *Contact* starring Jodie Foster.[64] Hulse and Taylor were caught up in the pulsar frenzy that had taken the early 1970s by storm, after Jocelyn Bell Burnell's discovery in 1967. At first they thought they had detected a normal pulsar, which pulsed with radio waves every 59 milliseconds (i.e. it rotates on its axis 17 times per second).

But as they continued to observe their newly discovered pulsar, they noticed something strange. The pulses weren't exactly 59 milliseconds apart – every time they took a measurement the time between pulses would be slightly longer or slightly shorter. That was weird: pulsars are some of the most precise clocks in the Universe; their period (the time between pulses) shouldn't change. When they plotted out the times that they were measuring, they got a wave shape: a sine curve. The variations in the time between pulses came back to the same measurements: every 7¾ hours. This was so regular that they could predict what the time between the pulses would be based on how long ago their last measurement was.

Hulse and Taylor realised that this could be explained if the pulsar was in orbit around another star,[65] with shorter measurements between pulses recorded as the pulsar moved

---

64 The Arecibo telescope sustained a lot of damage during Hurricane Maria in 2017. Two subsequent cable failures in August and November 2020 led to the telescope being safely decommissioned. Before work could start, though, the telescope collapsed and was damaged beyond repair.

65 Hulse and Taylor didn't realise the other star was another neutron star at the time, even though there was no visible companion. The nature of the other star was eventually confirmed by other groups of researchers studying the system.

towards Earth along its orbit, and longer measurements as it moved away from Earth along its orbit, repeating every 7¾ hours. This was the first time a pulsar had ever been discovered in a binary system like this, and so for the next six years, it was studied in excruciating detail until another curious property was spotted: the 7¾-hour orbit of the two stars was slowly decreasing. The orbits of the two stars were decaying; they were losing energy as they spiralled closer together. This energy was being lost to space itself and rippled away as gravitational waves.

It was Taylor, along with Lee Fowler and Australian astronomer Peter McCulloch, who published the results to the world in 1979, confirming that the decay in the orbits was exactly as Einstein had predicted (at least within the uncertainties of our knowledge on the distance of the pulsar from Earth), and not what other alternate theories of gravity being debated at the time had predicted.

This was the first indirect evidence for gravitational waves. Both Taylor and Hulse won the Nobel Prize in Physics in 1993 for their discovery of PSR B1913+16 which, according to the prize citation, was 'a discovery that has opened up new possibilities for the study of gravitation'.[66] Taylor, Hulse, Fowler and McCulloch weren't the first people to contemplate

---

66 Nobel Prizes can be shared by a maximum of three people. Lee Fowler unfortunately passed away in 1983, at the age of thirty-two, in a rock-climbing accident. I don't know why McCulloch wasn't also awarded the prize. Perhaps because the interpretation that the energy was being lost as gravitational waves wasn't quite agreed upon at the time the prize was awarded.

the existence of gravitational waves though; with the technological advancements in all other areas of astronomy charging forward post-Second World War, there were some who set their sights on actually detecting gravitational waves here on Earth. This reached a fever pitch in the 1970s, after a false claim of a detection of gravitational waves in 1969 by Joseph Weber, an engineer at the University of Maryland.

Weber had a large cylinder of aluminium that he claimed rang like a gong when impacted with a gravitational wave. Weber's supposed detections made no scientific sense, and were discredited by many leading astrophysicists at the time. But what his false claims did was spur on others to redouble the search and build their own gravitational wave detectors. The discovery of the orbital decay in PSR B1913+16 only added fuel to the fire. But how do you actually detect a gravitational wave?

Gravitational waves stretch and squash space itself as they move through it. So the distances between objects in space get shorter and longer as a wave passes by. If you can measure the distance between objects changing, then you can detect the presence of a gravitational wave. You need to do this very precisely though; the method of choice usually being to use a laser. A laser is a source of light of just one specific wavelength (and therefore the same colour, hence why you have the choice of a green or red laser pointer) that is emitted in the same direction to give a very tight beam. This means you can point it in one direction and know the majority of light will go in that direction, unlike a light bulb which emits light willy-nilly in all directions.

That means if you shine a laser at a mirror, the majority of the light will make it to the mirror and then reflect back off it, so you can still detect the same laser beam where it was first emitted (don't actually try this with a mirror at home folks; lasers can blind). Knowing the speed of light, you can then work out the round-trip distance the laser travelled thanks to the age-old classic equation: distance = speed × time. So there you have it, an accurate way to measure the distance between objects[67] (the laser and the mirror) to check if gravitational waves are passing by, squashing and stretching the distance between them.

The problem with gravitational waves, which Einstein himself pointed out, is that their effect (how much they squash and stretch) is absolutely tiny. We're talking a change in the distance between two objects of smaller than the diameter of a proton: less than 0.000000000000001 m. Measuring anything with that kind of precision, even with lasers, is a tall order. Instead, during the 1960s and 1970s (there's no real consensus on who had the idea 'first'), astrophysicists realised they could use a trick of physics to be able to measure with such precision; again due to the nature of lasers.

---

67 This is also how we precisely measure the distance from the Earth to the Moon. There are five 'retroreflectors' that have been left on the surface of the Moon, which are mirrors that make sure the light is bounced back in the same direction it came from (like a 'cat's eye' in the centre of a road at night). Three were left by NASA's Apollo missions and two by the Soviet Union's un-crewed Luna missions. With these retroreflectors and a very powerful laser, astrophysicists have been able to work out that the Moon is moving away from the Earth at around 4 centimetres per year.

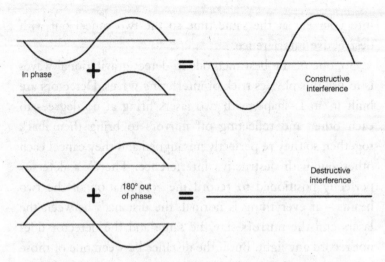

Constructive (top) and deconstructive (bottom) interference between waves that are in phase and out of phase.

The light given off by lasers is all the same. The peaks and troughs of every wave are lined up; physicists call that being *in phase* (like being in sync with someone). If you add a second laser into the mix, you can position it so that the waves in the two lasers are also in phase, so that when they meet at a detector the waves add together and you detect something twice as bright. We say that the two waves interfered with each other *constructively*. Or, you can position the second laser so that the waves are out of phase, misaligned so that they cancel each other out completely and no light is detected. In this case, we say the two light beams interfered with each other *destructively*. This is exactly how noise-cancelling headphones work – recording sound waves arriving at the headphones, and playing the inverted out-of-phase sound wave into your ear at the same time so the two cancel out with destructive interference.

So, one of the best methods to detect gravitational waves is to use this physics trick of interfering waves. Detectors are built in an L-shape, with two lasers firing at 90 degrees to each other and reflecting off mirrors to bring them back together, so they're perfectly misaligned and they cancel each other out with destructive interference. There's a detector perfectly positioned to record the combination of the two beams – if everything is normal, the distances between the lasers and the mirrors stay the same and the detector does not record any light. But if the distance between one of those laser-mirror pairs changes due to a passing gravitational wave, the phase difference between the two lasers changes and the detector will record a detection of some of the laser light. Depending on how bright the point is, between no detection to twice the brightness of a single laser, you can tell how much the two waves are out of phase as a fraction of the wavelength of the light emitted by the laser. This method (called *interferometry*, because you use the interference of the two waves) is how you can measure the tiny changes in distance between objects caused by gravitational waves, even down to less than the size of a proton.[68]

It was American physicist Robert L. Forward who built the very first prototype gravitational wave detector using

---

68 Gravitational wave detectors built in this way can therefore only detect a certain frequency of gravitational wave; it depends on the wavelength of laser you use, and the distance between the laser and mirror. It's nothing to do with the amplitude of the gravitational wave.

interferometry of lasers in 1971. Each L-shape was 8.5 metres long and the detector was left for 150 hours to record any gravitational waves, but was unsuccessful (it also didn't agree with any of Weber's gravitational-wave 'gong' detectors). It was American astrophysicist Rainer Weiss from the Massachusetts Institute of Technology (MIT) who pointed out that a much bigger distance than 8.5 m between the laser and the mirror would be needed to detect gravitational waves; in the early 1970s he calculated that to detect gravitational waves from the Crab Pulsar (formed in the supernova observed by Chinese astronomers in 1054) you would need 1 km (0.62 miles) between the laser and mirror. He even went as far as to suggest building an interferometer in space.[69]

It was in the summer of 1975 that Rainer Weiss met with his old friend Kip Thorne, an American theoretical physicist working at the California Institute of Technology (Caltech), known for his work on black holes and general relativity.[70] The two were attending a conference in Washington on cosmology and relativity and, according to Weiss, stayed up the entire night before the conference discussing the big unknowns in gravity research, before together deciding that their focus in the future should be on gravitational waves. In order to seriously tackle the problem they knew they'd need two things: 1) a whole lot of funding, and 2) an experimental

---

69 A dream that might come true in the twenty-first century thanks to NASA's plan for the Laser Interferometer Space Antenna (LISA), to be launched in 2037 (at the earliest).

70 He is notable in the twenty-first century for scientifically advising on Christopher Nolan's 2014 sci-fi epic *Interstellar*.

physicist (Weiss and Thorne were both very much immersed in the theory of gravitational waves, and not so much in the engineering or experiment design beyond prototypes).

Funding was not easy to come by: there were many problems to overcome, including a whole bunch of technological advancements. One thing that was needed was a way to isolate both the lasers and mirrors from any seismic activity. Although high-magnitude earthquakes that have the power to cause huge destruction are very rare, lower power earthquakes causing minor shakes, barely even noticed by us humans as we go about our daily lives, are very common. According to the Incorporated Research Institute of Seismology (IRIS), there are on average a few hundred earthquakes of less than 2 magnitude on the Richter scale (with about the same power as a lightning bolt hitting the ground) that occur every single day across the globe. Given the level of accuracy and sensitivity required of gravitational wave detectors, they needed to be isolated from this shaking, otherwise all you've built is a very expensive earthquake detector.

Similarly, even a heavy truck driving nearby could be enough to shake up the laser and mirror set-up. Burying the detector deep underground fixes the truck problem, but only exacerbates the seismic situation. It was Italian physicist Adalberto Giazotto, working at the University of Pisa, who cracked what to do. He was developing new suspension systems, which he called super-attenuators. He presented his new device at a meeting in Rome in 1985 and pointed out that they would be able to isolate the mirrors from any seismic activity. At the same meeting, a French physicist, Jean-Yves

Vinet, who had been working at the Applied Optics Laboratory in Paris (Laboratoire d'Optique Appliquée) presented his work on laser recycling, which allowed you to bump up the power of a laser so that it could still be detected over the large distances needed in gravitational wave detectors.

There was already great interest in Europe in building a gravitational wave interferometer, pioneered by French physicist Alain Brillet, but funding was once again proving to be the biggest barrier. Eventually, both the American and European collaborations were awarded funding (after losing out for many years prior to other projects, such as the Very Large Telescope, VLT, in the Atacama Desert in Chile[71]). The Caltech–MIT collaboration of Weiss and Thorne was funded by the USA's National Science Foundation (NSF) in 1988, and was dubbed the Laser Interferometer Gravitational-Wave Observatory (LIGO). The European collaboration of Brillet, Vinet and Giazotto was jointly funded by the French CNRS (Le Centre National de la Recherche Scientifique) in 1993 and in 1994 by the Italian INFN (Istituto Nazionale di Fisica Nucleare), and was dubbed the VIRGO observatory (after the largest nearby galaxy cluster, in the Virgo constellation).

Having solved their funding issues, the problem was now building and actually getting the interferometers working. LIGO suffered immensely in its early years due to disagreements between members on how to both build and manage the project. In 1994, an American experimental physicist called

---

71 A lot of my research field uses data from the VLT in Chile, so I'm rather grateful that this was also funded!

Barry Clark Barish was brought on as director of the collaboration. He was an expert in experimental high-energy physics and crucially had experience managing this new style of big-budget physics project. He redesigned the whole project, deciding that it would be built in two stages: an initial prototype, which could then be improved where needed to increase sensitivity and accuracy in the final stage. Given the intricacies of the interferometer method, this was a very smart move.

Construction of both LIGO and VIRGO prototypes progressed through the 1990s, with problems solved as they reared their heads. VIRGO found a site in Tuscany, Italy. LIGO was to be two separate detectors, one in Livingston, Louisiana and another in Hanford, Washington, both in the USA. Again, this was a smart move; it meant that if the detectors at both sites, separated by around 3,000 kilometres (1,865 miles), reported the exact same detection around 10 milliseconds apart (the travel time of light between the two sites) you could be sure that what you'd detected was a gravitational wave, and not a local disturbance (like a very heavy truck passing overhead). From the delay, you can get a good idea of a direction that the gravitational wave came from in space. Adding a third detector into the mix makes this even more accurate; you can literally triangulate the direction of the gravitational wave. So in 2007, the two separate projects of LIGO and VIRGO joined together to share results and detections.

Despite multiple observing runs through the late 2000s, no detections were made. Updates were needed to improve the sensitivity of the detectors and their isolation from seismic activity. These updates were made through the early 2010s and the detectors weren't switched on again until September 2015. In the days that followed, the detectors remained in 'engineering mode', so that tweaks and calibrations could still be made where necessary. It was during this time that an Italian astrophysicist, Marco Drago, a postdoctoral researcher[72] working at the Max Planck Institute for Gravitational Physics in Hanover, Germany, received an email from the automated LIGO system that a detection had been made at both the Livingstone and Hanford detectors.

The two detections were identical and looked the right shape to be a gravitational wave ripple from two black holes, but recorded at slightly different times at each detector with milliseconds difference. It could only be one of two things, either 1) a real gravitational wave, or 2) a fake model signal, artificially 'injected' into the system to check that all the procedures of detection were working properly. However, LIGO was still in engineering mode, meaning there was no way to inject fake signals yet. Drago knew this detection had to be real, but asked his colleague, another postdoctoral researcher called Andrew Lundgren, to double check. They called both Livingstone and Hanford to check whether there was anything unusual to report, but there wasn't. An hour after receiving the first email, Drago then sent an email to the entire LIGO collaboration asking if there was any way a spurious signal could be generated at both detectors, but got no reply. In the following days, senior LIGO members

---

72 Shout out to my fellow postdocs. We've got this.

confirmed to the collaboration that there had been no fake signals injected. Within two days of being switched on after its upgrade, LIGO had finally achieved what Weiss and Thorne discussed that night forty years earlier.

This discovery was perhaps the worst kept secret in the history of astronomy. The LIGO collaboration is so large that word eventually got out. I was doing my PhD at the University of Oxford at the time, and it felt like within a few weeks everyone in the astronomy community was abuzz with the news that LIGO had detected *something*. No one quite knew what had been detected until the news was officially announced at a press conference six months later, in February 2016. The entire collaboration had spent that time confirming that the signal had not been caused by a glitch in the detectors, earthquake or even spurious light sources. The signal, dubbed with the highly poetic name GW150914 (after the date it was detected on) was the first ever direct detection of gravitational waves, and its shape matched the predictions of Einstein's theory of general relativity for the spiral inwards and merger of a pair of black holes.

Not only was this another win for general relativity, but it was the first time in the history of humankind that we had observed the Universe with something other than light. We could 'see' in a whole new way. But it wasn't the visual graphs of the signal that captured the attention of the public; instead it was the sound produced when the signals were converted to frequencies in the human hearing range that delighted people worldwide. It's akin to the sound when you close your mouth around your index finger and push your finger into the inner part of your cheek and release it. Pop! It's perhaps one of my favourite parts of this whole story, that a cheeky pop can represent the most devastating of collisions between two of the Universe's biggest mysteries.

This discovery of gravitational waves was rewarded in 2017 with a Nobel Prize in Physics. The prize was split between Rainer Weiss, Kip Thorne and the savvy director of the LIGO team, Barry Barish. With the sheer size of the LIGO–VIRGO collaboration, and many other physics experiments worldwide, a prize awarded to just three people doesn't quite summarise the sheer scale of human effort that went into that one single discovery. There are over 1,200 people working in the LIGO collaboration alone.

This discovery of gravitational waves confirmed the existence of binary black hole systems – where two massive stars had previously lived, died and gone supernova – that astronomers had long suspected but had been unable to detect. It wasn't long before yet more detections were made, with another popping up in December 2015 before the first was even announced, and a total of fifty detections by October 2020. From black hole binary mergers, to black hole–neutron star mergers, and neutron star binary mergers. The neutron star binary mergers are often the most exciting detections, as we also detect a flash of light from these before their combined mass collapses into a black hole. This can give us an accurate distance to the pair along with a more accurate estimate of the Tolman–Oppenheimer–Volkoff limit for the maximum mass of a neutron star (or the minimum mass of a black hole).

We can't know just how many doors of discovery this

detection will open in the future, but what we do know is that it has completely and irrevocably changed the entire field of astronomy. In the same way that we were previously limited to only what our own eyes could reveal before telescopes gave us a way to view the whole spectrum of light, now gravitational wave detectors have given us a whole new way to see.

9

Your friendly neighbourhood black hole

In the wise words of Douglas Adams – DON'T PANIC. When I tell people my most fervent wish is for the Solar System to have a black hole of its own, they look at me with revulsion and horror. But as we've learnt previously, black holes are not hoovers – in the Solar System a black hole's role would be more one of a gravitational shepherd. So having a black hole in the Solar System wouldn't be a bad thing: it would be *cooooooool*.

Unfortunately there's been no confirmed reports (or 'sightings' – geddit?!) of a black hole in the Solar System yet. The closest known black hole to Earth is V616 Monocerotis, which may sound like some sort of disease, but is actually a black hole 6.6 times more massive than the Sun, squished into a space a touch smaller than the planet Neptune. It is fairly close to us at 3,000 light years away (about 28 million billion miles), but much further away than the nearest star to the Sun, which is only four light years away. So in the grand scheme of things, it's astronomically close but still not exactly what you or I would class as popping to the shops.

Thankfully, V616 Monocerotis is happily sat orbiting another star, one fairly similar to our Sun, which the black hole is slowly dragging material from into an accretion disk that gives off an occasional flare of X-rays, just to let us know that it's there. Apart from that, it doesn't have anything particularly remarkable about it, except for its proximity to Earth. And we've already agreed that there's nothing particularly special about our backwater region of the Milky Way.

The one thing that makes it truly special to the human race is not that we have detected light coming from the material spiralling around it, but that we have also sent a light signal towards it. On 15 June 2018, three months after the death of British astrophysicist Stephen Hawking, who devoted his life to understanding the mathematics of black holes, the European Space Agency sent a broadcast out in the direction of V616 Monocerotis in his honour. It'll arrive in the year 5475, and will be the first ever human 'communication' with a black hole.

V616 Monocerotis is only the closest *known* black hole, though. What if it's not truly the closest black hole? There could be another that's closer, perhaps a pair of black holes orbiting each other, like the system LIGO detected gravitational waves from, that don't have any material around them to heat up to flag to us with X-rays that they're there. Or perhaps even one hiding in plain sight in our very own Solar System?

That's not as mad as it first sounds, I promise. There's good reason to think that there might just be a tennis-ball-sized black hole hanging around on the edge of the Solar System, way out beyond the orbit of Pluto, *stirring sh*t up*. At first, the reason was that astronomers thought Uranus and Neptune's orbits were a little weird. So weird, that after Neptune was discovered in 1859 (after Le Verrier famously predicted where it would be), people immediately began searching for another planet ('Planet 9') beyond Neptune that could be disturbing both Uranus and Neptune's orbits: pulling on them due to gravity and making their orbits a lot more elliptical than those of other planets in the Solar System.

This elusive 'Planet 9' was finally thought to have been found in 1930, when, aged just twenty-four, American astronomer Clyde Tombaugh discovered Pluto. Tombaugh had taken up the mantle of searching for Pluto from fellow American astronomer Percival Lowell. Lowell was born into the Boston elite, and so of course studied at Harvard University. After graduating he ran a cotton mill in the city for six years and then decided to travel far and wide across Asia for the next decade. When he finally returned to the USA at the end of the nineteenth century, he decided to take up a career in astronomy. He went about it not as you or I would go about it, by applying for a job, but instead used his inherited and earned wealth to found a brand new observatory: the Lowell Observatory, just outside Flagstaff, Arizona, USA. Lowell specifically chose the location due to its high altitude and distance from city lights – the best conditions possible for astronomy – marking the first time the location of an observatory had been chosen in this way (rather than through convenience of location). It is now how the locations of all professional observatories are chosen, with the common themes of distance from populated areas, and high and dry climate. Think Mauna Kea, Hawai'i; the Atacama desert, Chile; or the Warrumbungle National Park, Australia.[73]

---

73 All of which are places us astronomers are more than happy to travel to, especially if we can tack on a few days holiday at the end of an observing trip.

It was at Flagstaff in 1906 that Lowell started a dedicated search for 'Planet 9' (or 'Planet X', as he referred to it). Like at Harvard College Observatory, where they were classifying stars, Lowell hired a team of women computers to do the tedious searching of photographic plates, which was headed up by Elizabeth Langdon Williams. Williams had just graduated with honours from MIT in 1903 with a degree in physics, becoming one of the first women ever to do so. She was initially hired by Lowell in 1905 to edit his scientific publications, before being asked to lead the team of computers at the observatory. Lowell had given Williams a rough idea of where he thought Pluto would be (orbiting in the same plane as Uranus, at about forty-seven times the Earth–Sun distance), and she was left to do the grunt work of calculating possible orbits for 'Planet 9' in order to recommend the regions of the sky that should be searched.

Lowell would then observe those areas of the sky with the telescope at the observatory frequently, comparing the most recent images with those taken previously to see if anything had moved in front of the background stars (again, by today's definitions, Williams was doing the astrophysics and Lowell the astronomy). He continued searching right up until his death in 1916, but never found what he was looking for. Although, with hindsight, we now know that the Lowell Observatory captured two very faint images of Pluto in 1915, but they were missed in the search.[74]

---

74 In 2000, Greg Buchwald, Michael Dimario and Walter Wild (three amateur astronomers) reported another 'pre-discovery' of Pluto in

After Lowell's death, the search halted for over a decade; in that time Williams married a British astronomer also based at the observatory, George Hall Hamilton, and was promptly dismissed from her position as lead computer because ~~early twentieth-century views on women were ridiculous~~ it apparently wasn't appropriate to employ a married woman. So, when the search finally resumed in 1929, it was the newly employed Clyde Tombaugh who took up the mantle. Tombaugh had impressed the observatory director, Vesto Melvin Slipher,[75] with the scientific drawings of Mars and Jupiter he'd made using a telescope he'd built and tested himself on his family's farm in Kansas.

---

photographic plates taken in August 1901 at the Yerkes Observatory in Williams Bay, Wisconsin. This is the earliest known 'pre-discovery', along with fourteen other observations of Pluto from observatories around the world. These extra observations are incredibly important to our understanding of Pluto's orbit. Pluto takes almost 248 Earth years to complete one orbit around the Sun, so has only moved along about 37 per cent of its orbit since its discovery in 1930. Extra observations back to 1901 take that to almost half of its orbit, allowing us to understand Pluto's orbit with greater precision.

75 Slipher was the first person to observe and record the redshifted light of galaxies in 1912, the first experimental evidence for the expansion of the Universe. Edwin Hubble is often wrongly credited for these observations; Hubble combined his own measurements of distances to galaxies with Slipher's observations of redshift to show there was a correlation between the two in 1929. It was George Lemaître who had predicted this correlation two years earlier (using Einstein's general relativity equations) and asserted that if this was the case then the Universe must be expanding. According to Allan Sandage (who used the correlation found by Hubble to derive the first accurate estimate for the age of the Universe in 1958), Hubble himself was always doubtful of the expansion-of-the-Universe interpretation of his results.

Tombaugh was given the rather tedious job of searching for 'Planet 9' by blinking back and forth between pairs of photographs of regions of the night sky taken a week apart. After a year of searching, he finally found an unknown object that had moved in images taken a few weeks prior, in January 1930. A few more observations confirmed the object was real and continuing to move in the same direction, and the discovery was finally announced to the world in March 1930.

The discovery made headlines around the world, and the question on everyone's lips was what to call the new planet in the Solar System. The Lowell Observatory had the right to name it, by virtue of discovering it, and they received over 1,000 suggestions through the post from eager astronomy lovers the world over. Constance Lowell, Percival's widow, who had taken over managing the estate, suggested Zeus (after the Greek God of the Sky), and even her and her husband's names: Percival and Constance. All of these were unsurprisingly dismissed by Slipher and Tombaugh (including Zeus, as all the other planets in the Solar System have Roman names, not Greek: Jupiter is Zeus's Roman equivalent).

Pluto is the Roman God of the Underworld, and according to Clyde Tombaugh, the name was originally proposed by an eleven-year-old in Oxford: Venetia Burney. This wasn't just any ordinary eleven-year-old though; this was the granddaughter of a retired librarian at the University of Oxford's Bodleian Library, Falconer Madan. Madan had friends in high places who he could relay the suggestion to, specifically the Savilian Professor of Astronomy and director of the University of Oxford's Radcliffe Observatory, Herbert Hall

Turner (remember, the author of *Modern Astronomy* from the prologue). Turner then sent a telegram to his colleagues at the Lowell Observatory, who included it on a shortlist of potential names (including Minerva and Cronus). A vote was held by the observatory staff, which was unanimous, and Lowell's 'Planet X' was officially named Pluto on 24 March 1930.[76]

In the end, Pluto was found just six degrees away from where Lowell (with Williams doing the calculation) predicted it would be. So at first, physicists were confident that Pluto was responsible for the oddities of Uranus and Neptune's orbits. Its mass was estimated based on how big it would need to be to affect them: seven times more massive than Earth. But due to how faint Pluto appeared (if it was bigger it would reflect more light and appear brighter) that mass was cast into doubt. By 1931, that estimate had been revised down to somewhere between 0.5–1.5 times the mass of the Earth, and it kept dropping through the twentieth century. Dutch astronomer Gerard Kuiper himself estimated it to be just 10 per cent of the Earth's mass in 1948, but this was still a gross overestimate.

It was in 1978 that Pluto's moon, Charon, was discovered by astronomers Robert Harrington and Jim Christy working at the United States Naval Observatory. From the orbit of

---

76 Most languages also use the name Pluto, with some using the literal translation for the 'God of the Underworld' in their own languages. For example, in Hindi, Pluto is known as Yama, after Yamarāja, the Hindu, Sikh and Buddhist deity of death and the underworld. Similarly, in Māori, Pluto is known as Whiro after Whiro-te-tipua, the embodiment of all evil who inhabits the underworld in Māori mythology.

Charon they were able to work out that the mass of Pluto was a measly 0.15 per cent of Earth's (that actually short changes Pluto a bit; modern estimates put it around 0.22 per cent of Earth's mass). This was far too small to account for the oddities of Uranus's orbit and it once again spurred searches for a planet beyond Pluto. These searches were brought to a halt by the results from the *Voyager* 2 flyby of Uranus in 1986 and Neptune in 1989 (the only craft ever to have visited either planet), which gave astronomers a more accurate estimate for both their orbits and masses. Taking all of these new measurements into account, the supposed weirdness to both of their orbits disappeared, along with the need for Lowell's supposed 'Planet X'. The fact that Lowell's predictions coincided with the area of sky where Tombaugh discovered Pluto is considered a happy coincidence.

What followed instead, throughout the rest of the twentieth century, was the discovery of many more small objects out beyond the orbit of Neptune in an area now known as the Kuiper Belt, after Gerard Kuiper. The Kuiper belt is an asteroid belt of sorts, but far larger (roughly twenty times wider) and more massive (up to 200 times more matter) than the asteroid belt between Mars and Jupiter. The search was kicked off by British-American astronomer David Jewitt and Vietnamese-American astronomer Jane Luu discovering the first two objects in the Kuiper Belt after Pluto in the early 1990s (1992 QB1 and 1993 FW). There are now over 2,000 known Kuiper Belt objects, but there are thought to be over a hundred thousand more small icy objects out there in the far reaches of the Solar System.

In 2005, three American astronomers working at the Palomar Observatory outside San Diego, California (Mike Brown, Chad Trujillo and David Rabinowitz) announced the discovery of a new object in the Kuiper Belt. At first this object was called 2003 UB313, but was eventually dubbed Eris (after the Greek goddess of strife and discord). Like Pluto, there are 'pre-discovery' images of Eris dating all the way back to 1954. Eris's moon was discovered a few months later, allowing Brown to calculate that Eris is 27 per cent more massive than Pluto. This made it the most massive object discovered in the Solar System since Neptune's moon Triton in 1846.

The world's press then dubbed it the 'tenth planet', but in astronomy circles that name was incredibly controversial. There were some in the community that thought Eris's discovery, along with other Kuiper Belt objects found at the same time, such as Makemake and Haumea, were the best argument for there only being eight planets in the Solar System – otherwise you would have more like fifty-three. Some astronomers started to argue that Pluto should be reclassified, but were wary of public reaction. When the Hayden Planetarium in New York displayed a model of the Solar System in 2000 with only eight planets, leaving Pluto off their model, it made headlines around the world because of the sheer volume of complaints that rolled in from visitors who were big Pluto fans.

Things finally came to a head in 2006, when at a meeting of the International Astronomical Union the official definition of a planet in the Solar System was decided upon by vote. A proposal for the definition had been put forward by committee and then members at the meeting were eligible to vote at a session chaired by none other than Jocelyn Bell Burnell (who discovered the first pulsar). The proposal passed the vote, and there are now three criteria needed to classify an object in the Solar System as a planet:

(i) It must be in orbit around the Sun

(ii) It must have achieved 'hydrostatic equilibrium' (i.e. it has enough mass that gravity has rounded it from a lumpy potato-shaped asteroid to something close to spherical)

(iii) It must have cleared the neighbourhood around its orbit

It's the third of those criteria that Pluto falls down on, along with all other objects of the Kuiper Belt, as they all inhabit the same neighbourhood of the Solar System.[77] Instead, they're classed as 'dwarf planets', along with a few other objects, like Ceres in the asteroid belt. It's safe to say the world did not react well to the decision. The American Dialect Society even chose 'plutoed' as its 2006 Word of the Year; *to*

---

77 Die-hard Pluto fans often complain that this definition should surely rule out the likes of Jupiter as well, since Jupiter has a collection of asteroids that have clumped together in front of and behind it in its orbit (known as the Trojan asteroids). But the mass difference between the goliath of Jupiter and a bunch of tiny asteroids is vast. Whereas the mass of the detritus of objects in the Kuiper Belt compared to Pluto is very similar. There's no comparison.

pluto something meant to demote or devalue it. I still don't think the internet is quiet over this demotion – anytime I bring it up it is met with absolute outrage. Although I like to point out to all those Pluto fans that now you can at least consider it 'King of the Dwarves'.

The study of these newly dubbed dwarf planets in the late 2000s unearthed yet more orbit peculiarities that couldn't be explained. For example, the dwarf planet Sedna has what's known as a 'detached' orbit. Unlike the other 'Trans-Neptunian Objects' of the Kuiper Belt (or TNOs), Sedna never crosses the orbit of Neptune; the orbits are elliptical, so you could say that its closest point to the Sun is still further away than Neptune's furthest point from the Sun (unlike Pluto and Eris, which do get closer to the Sun than Neptune's furthest point and were probably shepherded there by the gravity of Neptune during the Solar System's formation). Sedna actually orbits three times further out than Neptune on a highly elliptical orbit which takes over 11,000 Earth years. How did Sedna get into such a strange and distant orbit? One option is that it could be an object that was wandering interstellar space and was captured by the Sun. Another option is that it could've got pulled out there if the Sun had an interaction with a passing star, or, the most exciting option, by another massive planet on the edge of the Solar System.

It's this last idea that's favoured by the person who discovered Sedna: American astronomer Mike Brown (who also discovered Eris, leading to the demotion of Pluto that has earned Brown the nickname 'Pluto killer'). After the discovery of six more objects found with detached Sedna-like orbits at huge distances through the early 2010s, Brown and his Caltech colleague Konstantin Batygin (a Russian-American astronomer) investigated further. They found that not only did these objects have similar distances from the Sun, but they all orbited in the same plane, as if they had been shepherded there by an object on the far reaches of the Solar System. Brown and Batygin worked out that the most likely explanation was a planet somewhere between five to fifteen times more massive than Earth orbiting at the far edge of the Solar System.

Overnight, Brown and Batygin single-handedly sparked the search for yet another 'Planet 9' in the Solar System; but in the words of Carl Sagan, 'extraordinary claims require extraordinary evidence'. Planet 9 still remains a hypothetical planet, and despite many searches nothing has been found. One such search was done by volunteers on the Zooniverse citizen science online platform.[78] Similar to how Tombaugh found Pluto, volunteers were shown two infrared images from NASA's Wide-field Infrared Survey Explorer (WISE) mission which blinked back and forth so that they could spot if

---

78 There are many research projects that need help classifying huge amounts of data at https://www.zooniverse.org/, which has over 2.3 million volunteers worldwide. The Zooniverse started with the Galaxy Zoo project which was set up by British astrophysicist Chris Lintott at the University of Oxford to originally classify the 1 million galaxy images from the Sloan Digital Sky Survey. Chris also happened to be my PhD supervisor, and during that time I used data from Galaxy Zoo project to do 'big picture' galaxy evolution studies. My PhD was made possible by the efforts of 300,000 volunteers around the world classifying the shapes of galaxies, and I will be for ever grateful for their efforts. If you were one of those 300,000 – thank you.

anything had moved. While the project didn't find 'Planet 9', it did find 131 new brown dwarf stars beyond the Solar System and ruled out a huge area of sky for future Planet 9 searches.

What makes the search for Planet 9 so difficult is that, if it exists, it is estimated to orbit at a distance of over 500 times the Earth–Sun distance. That means it will take an incredibly long time for it to complete one orbit of the Sun, and so it's not expected to move that much on the sky on human time-frames. And so 'Planet 9' remains both hypothetical and elusive, with the orbits of the Sedna-like objects unexplained.

However, in 2020 a paper was published by Jakub Scholtz and James Unwin linking not only this unexplained phenomenon but another that, at first, appears completely disconnected. The Optical Gravitational Lensing Experiment (OGLE) run by the University of Warsaw uses a telescope in the Atacama Desert in Chile to spot if anything changes brightness in the sky. That can be anything from pulsing stars, to supernovae, or something called a *microlensing* event. This is when a compact object, like a neutron star or black hole, passes in front of a background star. The light of the background star gets bent as it travels along the curved space around the compact object, which acts like a lens to briefly brighten the background star. From how much the background star changes in brightness, and how long the change lasts, you can work out, once again using Einstein's general relativity equations, how massive the compact object doing the lensing is.

The OGLE survey has been running since 1992 and in that time has spotted many a gravitational lens caused by black holes in the Milky Way, all formed when a star went supernova, giving a black hole anywhere above the Tolman–Oppenheimer–Volkoff limit of around three times the mass of the Sun. But the OGLE team also reported that they'd observed six ultra-short micro-lensing events in the direction of the centre of the Milky Way (which also crosses the plane of the Solar System) that had to be caused by objects just 0.5–20 times the mass of the *Earth*. Such a low mass meant that either this had to be a population of rogue, free-floating planets that had been ejected from whichever star system they formed in, or a population of *primordial* black holes. A primordial black hole is a hypothetical type of black hole that formed in the very early Universe when the Universe was much denser; if real, they'd be the oldest black holes in existence. Theoretically, if enough matter happened to clump together randomly during that time, a tiny black hole could form, an idea that was developed by Stephen Hawking in the 1970s.

What Scholtz and Unwin pointed out in their paper, entitled 'What if Planet 9 is a Primordial Black Hole?',[79] was that the two mass ranges, as predicted by Brown and Batygin for Planet 9 (5–15 times Earth mass) and seen by the OGLE team (0.5–20 times Earth mass) were remarkably similar, and perhaps one could help explain the other. Perhaps Planet 9 was once part of this population of objects causing the micro-lensing events seen by OGLE; either a captured free-floating planet or a captured primordial black hole.

Capture of Planet 9 is just one possible explanation for how

---

79 A title which is practically the scientific equivalent of clickbait: I don't think I've ever clicked on a newly published paper so fast.

the hypothetically rather large Planet 9 might have formed on the edge of the Solar System. Other options are 1) that it somehow managed to form where it currently orbits or 2) that it formed further in, closer to the Sun, and then migrated outwards. That first option is unlikely, as it's not very dense at the edge of the Solar System, so 4.5 billion years is still not enough time to bring together all of those far-flung tiny clumps of rock to form a planet that large. The second option is also problematic, because you need an event that kick-started the migration but also one that stops it in its current orbit, which could perhaps be an interaction with a passing star again, but that seems unlikely. So with those two ideas out, the hypothesis of a captured Planet 9 is currently the most favoured.

Planet-system formation models have shown that during the chaos of planet formation around stars, with bits of rock colliding and clumping together under gravity or perhaps slingshotting around each other, many planetesimals (i.e. baby planets) get flung out of the melee into interstellar space. We think one such object, dubbed 'Oumuamua', travelled through the Solar System back in 2017, passing within just 24,200,000 km of us here on Earth. That's about 15,040,000 miles, or about 16 per cent of the Earth–Sun distance. Given how big space is (think about the huge distances involved and then remember that space is three-dimensional, so you need to cube whatever huge number you just thought of) we think these events are incredibly rare, and the gravitational capture of such an object by the Sun even rarer. However, that likelihood of capture doesn't change whether the object is a rocky planet or an incredibly dense primordial black hole.

The beauty of this hypothesis – that Planet 9 is a black hole – is that it also explains why we haven't found it. Not just in recent searches, like with the Zooniverse, but in previous searches throughout the last few decades which found other Kuiper Belt objects. Not only would we not get any light from a black hole but nothing would ever get close enough to be directly impacted by it. If the black hole Planet 9 turned out to be five times the mass of the Earth, its event horizon would be just 9 cm across; about the size of a tennis ball.

A circle 9 cm across. This could be the size of a primordial black hole five times the mass of the Earth lurking at the edge of the Solar System.

Now, as much as I desperately want this hypothesis to be true, the problem with Planet 9 turning out to be a primordial black hole would mean that it would be incredibly difficult to find evidence for it. Although, if it was a black hole that has existed since the very early days of the Universe, in the past 13 billion years or so, it will have collected a little halo of matter around it. Not in an accretion disk necessarily, just a clump of matter that is shepherded along by its movement through space, making the region around it much denser than normal. If that area is denser, it increases the chances that some of that matter will encounter some very rare anti-matter. Thankfully, we have a lot more matter in the Universe than anti-matter, otherwise nothing you have ever seen in your life, including the stars themselves, would ever have existed. Because when matter meets anti-matter it turns back into pure energy released as gamma rays; the most energetic type of light.

So, if the Solar System has its very own pet black hole, we should be able to detect that radiation with the gamma-ray telescopes that we currently have in orbit around Earth. The Planet 9 search isn't just one that optical and infrared astronomers are involved with, the gamma ray astronomers have now been caught up in the furore as well. Because the idea that our nearest black hole could be light hours away rather than light years is enough to capture the heart of even the most hard-headed astrophysicist. To me, the theoretical evidence is very compelling, but perhaps I'm slightly biased as a black hole scientist; a black hole right on my doorstep would be the best present the Universe could ever get me.

10

Supermassive-size Me

Super-massive-size Me

One phrase I find myself saying nearly every day is: 'at the centre of every galaxy there's a supermassive black hole'. I say it so casually. It's a throwaway remark; like the sky is blue, the Earth is round, or Taylor Swift is the greatest lyricist of my generation.[80] I take for granted that it's a fact humanity knows. But rewind just fifty years and that phrase would have been met with disbelief and perhaps even a guffaw or two from a fellow physicist. That change in attitude didn't happen overnight. It took decades; another reminder that scientific theories don't just spring out of the ground fully formed; they take time.[81] Scientists are working with a few tiny scrap pieces of a jigsaw puzzle that didn't come with a picture on the lid, and so they have no map of what they're working towards. As more pieces of evidence are collected, the big picture starts to take shape; pieces that didn't look like they were connected at the beginning turn out to fit together and an accepted theory emerges.

The very first piece of the supermassive black hole jigsaw puzzle came in 1909. A chap named Edward Fath was observing

---

80 'So casually cruel in the name of being honest.' My fellow Swifties know.
81 As Gimli so eloquently puts it about dwarves in Tolkien's *The Lord of the Rings*.

'spiral nebulae' at the Lick Observatory, just outside San Jose, California.[82] Back then, 'nebula' was the term used to describe anything that didn't look like a star in the sky. It encompassed all the dusty, fuzzy things in the sky (the word 'nebula' is Latin and literally means 'mist' or 'cloud'). Back in 1909, the size of the Universe was thought to be as big as the Milky Way – the most distant thing known was a star a hundred thousand light years or so away on its edge. All the nebulae were therefore also thought to be inside the Milky Way; either they were places where new stars were being born out of giant gas clouds, or they were the remnants of a star that had gone supernova and dispersed all its outer layers back into space.

---

82 I've been lucky enough to observe at the same observatory. I was very excited about the trip as observatory locations are obviously chosen for their very clear dark skies. I had plans to lay myself down on a blanket outside and gaze at the stars in the warm Californian night air while the telescope was taking the thirty-minute exposure images of the galaxies I was studying. I arrived at the observatory to find signs everywhere telling me to beware of the mountain lions. The observatory staff told me not to worry, they were rare and only ever seen when their prey were around: deer. On the first night I decided to be brave and head outside to see the stars, but after five minutes of incessant nervous glancing at the dark trees around me I heard a rustle and saw three deer bound out of the tree line in the starlight. It was a sight that should have taken my breath away. Indeed, it did leave me gasping for breath, as I turned tail and sprinted back up the steps of the telescope building to get away from the mountain lion I was convinced was about to follow them. I spent the remainder of the trip cloistered indoors until the observatory staff told me about the balcony running around the edge of the telescope dome. After convincing myself that a mountain lion couldn't possibly jump that high, I finally found the perfect spot to sit down, kick my feet over the edge and lean back to gaze at the stars.

Splitting the light from a nebula through one of Fraunhofer's spectrographs reveals the unique fingerprint of light for that object. By doing this, you can tell what the nebula you're looking at is made of. But unlike what we see for stars, where there are gaps of specific colours (i.e. specific wavelengths), for gas clouds like nebulae there are instead extra bright patches where we'd usually see those gaps. Instead of absorption of light by different elements, we're seeing emission (like Kirchoff and Bunsen saw when they burnt sulphur).

As we learnt in Chapter 5, Niels Bohr explained how every electron orbits at a very specific distance from the nucleus, crucially, with only a limited number of special positions where an electron can orbit to keep the atom happy and stable. The orbit position of the electron tells us about how much energy it has to keep it in that orbit. This means the electrons around atoms in specific orbits have very specific amounts of energy. If you give an electron more energy though, perhaps by shining ultraviolet light on it from a nearby star, you can cause an electron to jump to a new position, giving it enough energy to jump to the next stable orbit (this is the absorption that happens in stars; with enough energy the electron escapes the atom entirely and becomes ionised). We say that the electron is in an 'excited state'; like a teenager on their first caffeine high.

Electrons aren't supposed to be in excited states in atoms though, because like teenagers, they like to be at the lowest energy possible. So, as soon as they can, the electrons lose energy to drop back to their original orbit. It's always the exact same amount of energy that the electron loses, because remember, there are only certain positions where the electron can orbit the atom to be happy and stable. That energy is lost as light. Since the same amount of energy is lost each time, the same wavelength of light is emitted, and so the same colour is always given off. So hydrogen gives off a lot of light at a specific wavelength of 656.28 nanometres, which is a deep red colour. When we split the light from a large cloud of glowing hydrogen gas through a prism into its rainbow of colours and trace the amount of light received of each colour, we get a huge peak of red-coloured light at 656.28 nm, which resembles a stalactite in shape.

Spotting the stalactite peaks of different colours on the traces from spectrographs that indicate when a specific element is present is key to understanding what kind of nebula you're observing. If there's lots of hydrogen then it's likely that you're looking at a nebula where new stars are being born, or if there's oxygen, carbon and nitrogen colours then you're looking at a nebula where a star has died: the supernova poop.

Anyway, back to our chap Fath in 1909. He was on the hunt for signatures of either supernova poop or of pure hydrogen gas in the light coming from a different type of nebula – the 'spiral nebulae'. What he found was that the spiral nebulae didn't fall into either one of those categories; instead they looked like the traces seen when observing clusters of stars with *both* the signatures of hydrogen and the heavier elements (and also some absorption of light too). What Fath had observed, but didn't realise at the time, were galaxies; islands in the Universe made of billions of stars. Just like our own Milky Way. This was the first of many results that contributed to the jigsaw puzzle for the size of our Universe. It was only after the work of scientists like Henrietta Leavitt, Heber Curtis and Edwin Hubble throughout the first two decades of the twentieth century that the distance to these 'spiral nebulae' could be measured. It was then that the scientific community finally realised that the Universe was far larger than they ever considered before: the Milky Way was no longer the only kid on the block.

With this mind-blowing realisation, one of Fath's other observations went largely ignored: one of the traces from the 'nebulae' that Fath had observed looked different again from all the others. It had signatures of hydrogen, oxygen and nitrogen but they were much stronger and brighter than had ever been seen before, as if there was an extra source of energy causing them to glow. So not only had Fath observed galaxies without knowing it, he had also unwittingly observed the gas glowing as it spiralled around what we would one day call a supermassive black hole. Of course, it would be decades before anybody recognised what Fath had really observed. This is often dubbed an unknown known – the things that we've observed or done the experiment for, but have missed the meaning behind. It fascinates me to think about all the experiments that have been done in the past few decades that have likely already revealed something extraordinary, but we still don't have the knowledge to understand what else they might be telling us. Or perhaps, even more likely in the era of data science and 'big data', information that's buried somewhere on a computer archive that has been missed by human eyes.

Similarly, Fath's strange observation of a galaxy with a very different trace of light was largely forgotten while astronomers and astrophysicists alike got distracted by what were considered the 'bigger questions' for a few decades. After determining that the Milky Way wasn't the entire Universe in 1920, their focus turned to how the Universe began. This continued for much of the interwar years, eventually leading to the development of the Big Bang Theory for how the Universe has evolved and expanded over the past 13.8 billion years. A worthy pursuit, but perhaps delaying our knowledge of black holes for a few decades. It wasn't until 1943 that American astronomer Carl Seyfert finally picked up Fath's observations and once again observed six galaxies with similar-looking traces of light. What he noticed was that the emission of light from hydrogen gas in these galaxies wasn't a sharp peak – instead it was smeared out into something that looked less like a stalactite and more bell-shaped.

Seyfert guessed that this smearing was due to Doppler shift; the light was being stretched and squashed as it moved both away from and towards us. If the glowing hydrogen gas in a galaxy is orbiting something, then some of the gas will be moving towards you and the light emitted will be squished to a shorter wavelength than was first emitted by the electrons jumping orbits, and some of the gas will be moving away from you and the light emitted will be stretched out to a longer wavelength. This is what turns our nice stalactite shape into a broadened-out bell shape. But here's where it gets really clever – the amount of broadening is related to how fast the hydrogen gas is moving. And if you know how fast the gas is moving, then you can work out how massive the thing is that it's orbiting.[83]

The Doppler shifts that Seyfert measured for his six galaxies were *huge*. Unprecedentedly large. At this point you might think people would have started to realise that there had to be a massive object somewhere in these galaxies in order to create this kind of smeared-out trace of light. But again, people didn't yet have all the knowledge needed to understand what Seyfert had observed. It would be another twenty years (with the work of Stephen Hawking and Roger Penrose in the late 1960s) before theoretical physicists even began to take the idea of black holes seriously.

Seyfert's work wasn't the only new result found in the post-war era. During the Second World War, the need to pick up faint radio signals from afar resulted in huge leaps forward in radio technology. After the war, those antennae were turned towards the sky, and many observatories with telescopes detecting radio waves were set up around the world, from Manchester[84] and Cambridge (where Hewish and Bell Burnell

---

83 This is exactly how I measured the masses of supermassive black holes in the centre of some galaxies during my PhD, after observing them with a telescope in the Canary Islands on the island of La Palma. It blows my mind not only that was I able to do that as part of my job, but also that we as humans are capable of it. The fact that collectively we have been able to piece together all the scraps of knowledge from chemistry, quantum physics and astrophysics to be able to measure the masses of supermassive black holes billions of light years away is something I will never get over, no matter how many times in my career I might do it.

84 You may think of Manchester as a terrible place to put a telescope, considering it's one of the rainiest cities in England (damn that relief

were discovering pulsars) in the UK, to the outskirts of Sydney, Australia. Instead of being used to pick up radio signals on Earth, bigger and bigger antennae were built to pick up even fainter radio signals from space, and radio astronomy was born.

It was the efforts of radio astronomers in cataloguing the new objects they were detecting in the sky that gave us a fair few more pieces of the jigsaw puzzle. First, one of the strongest radio signals in the sky was detected coming from the direction of the constellation known as Sagittarius. The father of radio astronomy himself, Karl Jansky, had detected radio emission coming from the direction of Sagittarius way back in 1931, but it fell to two Australian astronomers, Jack Piddington and Harry Minnett, working a radio telescope in Potts Hill, Sydney in 1951, to resolve that radio emission to a bright point in the direction of the centre of the Milky Way (astronomers had already agreed that the direction of the centre of our galaxy, the Milky Way, was in the constellation of Sagittarius since more stars could be seen in that direction – like looking towards

---

rainfall over the Pennines – the rainclouds from the Atlantic hit the barrier of the Pennine Hills running down the middle of England and abruptly stop, resulting in them dumping out all the water they picked up from the Atlantic over the North West – a phenomenon I am all too familiar with after growing up in Chorley, Lancashire). But that's the beauty of radio astronomy: you don't need clear skies to do it. Radio waves easily go through clouds, otherwise we'd never get any signal on our favourite radio stations on overcast or rainy days. Hell, you can even observe during the day with a radio telescope if you're clever about it – although it's still a good rule of thumb not to point a radio telescope at the Sun, since they're designed to focus tiny scraps of light, not the telescope-melting amount of light from the Sun.

the city centre and seeing more lights than if you look out towards the suburbs[85]). The second thing they found was a large number of radio-emitting objects scattered across the sky in all directions that didn't coincide with any object that had been seen by visible light. That made people wonder whether the objects producing these radio waves were so far away that the visible light from them was simply too faint to see with the optical telescopes available at the time.

Along with radio astronomy, X-ray astronomy was also on the rise after the Second World War, literally, with the use of balloons and rockets. Giacconi had discovered Scorpius X-1, as we learnt in Chapter 7, and Iosif Shklovsky explained Scorpius X-1 through accretion of material around black holes (and neutron stars) a bit heavier than the Sun found in our own Milky Way. But as X-ray astronomy gained in popularity, people started to spot other X-ray sources peppered across the sky that were incredibly faint, and yet incredibly energetic. To explain the incredibly energetic X-rays seen from these very faint, unknown sources (dubbed 'quasars' – quasi-stellar objects), you would need accretion around an unfathomably large object. It was British astrophysicist Donald Lynden-Bell[86]

---

85 Figuring out the shape of the Milky Way wasn't an easy task for astronomers either, because we're stuck inside of it. Imagine trying to make a map of your city without being able to leave your house!

86 Lynden-Bell is another BNIP (Big Name in Physics), who served as President of the Royal Astronomical Society and as the first director of the Institute of Astronomy at Cambridge University, when it formed from the merger of Hoyle's Institute for Theoretical Astronomy and the Cambridge Observatories in 1972.

who, in 1969, first proposed the idea that the huge amounts of energy coming from quasars could be explained by accretion onto an incredibly large collapsed object (much larger than the one powering Scorpius X-1 in the Milky Way), and suggested that all galaxy centres could have collapsed in this way. He even suggested our own galaxy, the Milky Way, could be a 'dead quasar' (i.e. a collapsed object that was no longer accreting material).

It was the Hubble Space Telescope, launched in 1990, that eventually detected visible light from these X-ray and radio sources dotting the sky, confirming that they were indeed incredibly distant galaxies. These incredible distances meant that the X-rays and radio waves were even brighter than first thought, far too bright to be caused by accretion onto a black hole just a few times more massive than the Sun. In fact, when they corrected for those immense distances, astronomers found they were even brighter than the very faint X-rays observed coming from the centre of the Milky Way. The logical conclusion was that there must not only be accretion onto an incredibly massive object occurring in these distant galaxies, but also in our own. Since we couldn't see any such object towards the centre of the Milky Way, these objects eventually got dubbed MDOs – Massive Dark Objects – in part because people were incredulous over the idea of a black hole so large, so *supermassive*, you might say.

During the 1990s, interest in what was going on in the centre of the Milky Way spiked. The problem was that seeing to the centre of the Milky Way is extremely frustrating because there's lots of dust and stars in the way, blocking the view. All hope was not lost, however: this was infrared astronomy's time to shine. Infrared light has a longer wavelength than visible light, meaning it easily passes around much smaller dust particles and allows us to see through to the centre of the galaxy. This technology kick-started a decade-long experiment to observe the positions of the stars in the very centre of the Milky Way, led by American astrophysicist Andrea Ghez at the University of California, Los Angeles and using the Keck telescopes on Mauna Kea in Hawai'i.[87] Ghez and her team recorded how the positions of the stars changed in order to determine their precise orbits around the centre. This is the same thing we do when we spot asteroids in the Solar System;

87 Mauna Kea is another place that I have been fortunate enough to visit in my time as an astronomer. I spent six days observing with the Caltech Sub-millimetre Telescope (affectionately referred to as the golf ball) and then two days snorkelling at sea level (if I hadn't become an astrophysicist I'd be a marine biologist). Mauna Kea is 4,207 metres tall (13,800 feet) and so altitude sickness really starts to kick in. Falling asleep at night (or during the day since you observe during the night and sleep all day on an observing trip) is nigh on impossible because your body constantly thinks you're not getting enough oxygen due to the thin air. You know that feeling when you're going to sleep and you jerk awake because you thought you were falling? Turns out your body does that when it's short on oxygen too (it's known as a myoclonic jerk). When I made it back down to sea level I slept for fifteen hours straight. The lack of oxygen at that altitude also affects your eyes, so when you step outside of the telescope building to look at the stars, you find you can't see as many as you thought because your brain has redirected the precious oxygen it can get to your internal organs. Breathing in from an oxygen canister causes a practical explosion of lights in front of your eyes as thousands of fainter stars come into view. It's magical. But probably not recommended by health and safety.

we observe how their position changes night on night, and then use that to figure out their orbit around the Sun. By studying the orbits of the stars in the centre of the Milky Way we can also determine how massive the object is that they're orbiting. We've now even seen one star complete an entire orbit around the centre in just sixteen years at a speed of over 11 million miles per hour. Compare that with the 250 million years it takes the Sun to orbit around the centre at 'only' 450,000 miles per hour.

In 2002, the results of Ghez's project were published, and astronomers finally knew how massive the dark object at the centre of our galaxy was: four million times the mass of the Sun. It is found in an area sixteen times the distance between the Earth and the Sun (to put that into context: Uranus orbits at nineteen times the distance between the Earth and the Sun[88]). For something to be so big in such a relatively small space and invisible to all wavelengths of light, there was only one thing it could possibly be – a supermassive black hole.[89] Proving this won Andrea Ghez the Nobel Prize in Physics in 2020, shared with German astrophysicist Reinhard Genzel, who had the first crack at using the orbits of stars to study the object at the

88 This is the size of the region inside the orbit of the closest star to the centre. The actual event horizon of the supermassive black hole is just seventeen times bigger than the Sun's diameter.

89 However, there was still a lot of debate between astronomers – as there had been since the early 1990s – about whether it was one single black hole or a swarm of black holes. It is in fact one supermassive black hole, because a swarm would be completely unstable with black holes flying off in all sorts of directions. But if I'm being honest, I'm sort of disappointed a swarm of black holes doesn't exist!

centre of the Milky Way, and with British mathematician Roger Penrose for his work with Stephen Hawking in the 1960s showing that black holes were inevitable in nature.

Accretion of gas onto a supermassive black hole explains all the X-ray and radio observations astronomers puzzled over during the twentieth century. The supermassive black holes at the centre of distant galaxies were so massive that the superheated gas spiralling around them was unbelievably hot, and so gave off unbelievably energetic X-rays. Heating gas to these extreme temperatures means that even the very atoms themselves separate into their constituent particles, so that the electrons are no longer bound in orbits around the nucleus. This means you have charged particles moving through space, which, when they move through a magnetic field, give off radio waves. This accretion onto a supermassive black hole, the idea used to explain all these observations and complete this scientific jigsaw puzzle, has eventually become known as 'active galactic nuclei unification theory'. To me, it once again represents one of the most misunderstood concepts about black holes; they're not 'black', they are the brightest things in the entire Universe. Completely unmissable, blazingly bright mountains of matter.

We're now lucky enough to even have an image of that superheated material spiralling around a black hole in the famous 'orange donut' picture: this is the very first image ever taken of a black hole, specifically the one in the centre of the nearby galaxy Messier 87. The orange light shown in the picture shows the radio waves detected from the disk of material spiralling around the black hole. There's an ominous shadow cast on this orange glow from the black hole, from which no light can escape. Compare that shadow of black in the centre to the dark on the outskirts of the image. No light can reach us from the inside because it is one of the Universe's heaviest, densest objects, surrounded by hot, furious activity. Whereas on the outside, there is no light because it is the quietest, coldest, emptiest place in the very same Universe. I get chills every time I look at it.

The first ever image of a black hole, taken in radio waves by the Event Horizon Telescope in 2019, of Messier 87*.

11

Black holes don't suck

11.

Black holes aren't...

When people try to picture black holes, they can't help but imagine them as the hoovers of the Universe; pulling in and gobbling up everything and anything around them. But that couldn't be further from the truth, because black holes don't suck.

Think about the Solar System: 99.8 per cent of all the matter in the Solar System can be found in the very centre, in the Sun. It completely dominates the Solar System and is utterly massive in comparison to everything else. Even Jupiter, the 'King of the Planets',[90] is only 0.09 per cent of all of the mass in the Solar System. Earth is a piddly 0.0003 per cent of the share. Despite the Sun's gravity dominating in this way, all the other inhabitants of the Solar System, from planets to asteroids and comets, happily orbit the Sun without 'falling in'. As general relativity explains, the Sun curves space and the planets travel along that curved space. To get the Earth to move closer to the Sun, you would somehow need to take away some of the Earth's energy, to disrupt the perfect gravitational balance it's currently in.

The region around black holes is exactly the same. Sure,

---

90 King is a disputed title in this house. Saturn is my personal favourite.

black holes are massive, but their dimensions are relatively tiny. Remember that if we collapsed the Sun down to a black hole the Schwarzschild radius would only be 2.9 km. Let's imagine for a minute we could make that happen; at first we'd probably notice that someone had turned the lights out, but apart from that we wouldn't notice a thing. Earth's orbit wouldn't change at all because the thing that it's orbiting around hasn't changed in mass, and the distance of Earth to the Sun hasn't changed, so the pull of gravity would be exactly the same.

But anything too close to that roughly 6 kilometre-across Sun-turned-black hole probably wouldn't be so lucky. The curvature of space near it would be dramatic, increasing the force exponentially. Anything further away, though, would just continue to orbit this theoretical black hole, forever tracing the same path through space in an endless loop. This is why when I say we are all orbiting a black hole at the centre of the Milky Way, there is no need to panic. Unless you spend your days terrified over the possibility that the Earth is going to fall inwards towards the Sun, then you can sleep soundly knowing that the Milky Way's black hole is merely shepherding the Solar System around the Milky Way. The Solar System is not spiralling in to the centre. It's on a very happy orbit; there's no doomsday scenario at the end of time where we fall into a black hole.

In fact, it's incredibly rare that *anything* makes it into black holes at all. It's a wonder that some of them have managed to get so supermassive. Take the black hole at the centre of the Milky Way; at about 4 million times the mass of the Sun, it has an event horizon just seventeen times bigger than the Sun's diameter. Just sit with that for a second; 4 million times the amount of matter found in the Sun, fitting well inside the orbit of Mercury. You'd think a behemoth like that wouldn't struggle with accreting any matter that got too close, but that's exactly what happened in early 2014.

Back in 2002, in the same year that the paper from Andrea Ghez's group confirmed that the only thing the centre of the galaxy could be was a supermassive black hole, something a little weird-looking was spotted on images of the centre of the Milky Way. It turned out to be a gas cloud, and by 2012 people had worked out that it was most definitely on its way towards the danger zone around the Milky Way's supermassive black hole. This was a once-in-a-lifetime chance for astronomers, because in the words of Douglas Adams: 'Space is big. You just won't believe how vastly, hugely, mind-bogglingly big it is.'[91]

Astronomers don't really do experiments, as such. The entire Universe is our experiment and we observe it in different ways at different times and watch how it changes. It means that if you want to know how matter behaves when it gets too close to a black hole, you can't just set that experiment up and make it happen. You have two options, either 1) simulate it happening on a computer and hope you didn't miss any laws of physics or 2) wait around billions of years for it to occur. The fact that this gas cloud, dubbed 'G2', was going to come within spitting distance of the supermassive black

---

91 From *The Hitchhiker's Guide to the Galaxy*.

hole at the centre of the Milky Way wasn't just once-in-a-lifetime, it was a once-in-a-billion-years type of opportunity.

So, as the gas cloud was torn apart slowly over the next two years, the astronomy world held its breath, and by 2014 fireworks were expected! But instead, astronomers got more of a flop. It was Andrea Ghez's group, using the Keck telescopes once again, that confirmed the G2 gas cloud was still intact. The gas cloud had looped around the centre of the galaxy relatively unscathed, despite passing as close as thirty-six light-*hours* from the black hole (about 2,375 times the size of the event horizon). Perhaps a star held it together against the pull of the black hole's gravity? Who knows. But what this serves to demonstrate is that black holes aren't just endless hoovers sucking material in. This gas cloud got as close as we'd ever seen something get before and it still didn't 'fall in' to become part of the black hole. Sure, it got a bit beaten up – it looks less like a cloud and more like an aeroplane contrail now – but it lived to fight another day, or at least drift around space indefinitely.

I can't help but anthropomorphise things in space when I think about events like this. I picture the G2 gas cloud streaking away from the black hole thinking *phew!* and warning every other gas cloud they happen upon not to go near the scary elephant graveyard[92] at the centre of the galaxy. The story of G2 gets passed around and used as a cautionary tale for millennia by parents of little gas clouds: 'Have a nice day,

---

92 Thanks to *The Lion King*, an elephant graveyard is always the scariest thing I can think of.

love; don't stray too close to the black hole! You don't want to end up like G2!'

And yet despite G2's escape, some gas clouds still end up under a black hole's control. We see this in the accretion disks around much more active supermassive black holes in the centres of other galaxies. Accretion disks are made of material that has made its way to the centre of a galaxy and not been as lucky as G2. Instead, it has been captured in orbit around the supermassive black hole. But as we just reasoned with planets around the Sun, material in orbit is not in danger of being 'sucked in' to the black hole. It will happily continue to orbit unless it loses energy in some way.

An accretion disk is an incredibly dense place. There's a huge amount of gas moving at immense speeds. Collisions between particles, like atomic nuclei (having separated from their electrons and become a plasma because it's so hot), are very common. These collisions are akin to those between balls in a game of pool. You give the white cue ball some energy by hitting it with the cue, and it then impacts with another ball, transferring that energy. Sometimes in those collisions the cue ball, if hit just right, will stop on impact, losing nearly all of its energy, and sometimes it will travel on with the other ball with a fraction of the energy it had before.

The same thing can happen to the particles in accretion disks; random collisions can transfer energy, imparting some particles with more energy so that they can move away from the black hole, and stealing energy away from others so that their orbit decreases. Enough of these random collisions can eventually strip a gas particle of enough energy so that it crosses the region around the black hole where you can have a stable orbit, and tumbles in beyond the event horizon to add to the mass of the black hole. Finally, one single particle has been accreted.

It can take over 500 million years for a supermassive black hole to accrete just half of all the matter in its accretion disk in this way as there's a limit to how fast this process can occur around a black hole. Rather ironically, the limit is named after Arthur Eddington (who we met earlier), who doggedly argued against the existence of black holes for so long. To be fair, it's a concept that doesn't just apply to black holes, but to all things that glow, including all the stars in the Universe.

Eddington had always been focused on stars and their interiors. How were they powering themselves? How much power they did they produce? To answer these questions he started by focusing on how stars stopped themselves from collapsing. Like Kelvin, Eddington reasoned that for stars to be stable spheres that didn't pulse in any way, the force of gravity crushing inwards must be balanced by the amount of energy released inside by whatever process powered stars. Since stars are hot, most astronomers assumed that it must be thermal energy alone pushing outwards, but Eddington added something extra: radiation pressure. Stars aren't just hot, they also shine, giving out huge amounts of light which exert a pressure outwards, resisting the crush of gravity.

When light hits things it can transfer energy. In theory, if you could build a laser powerful enough, you could use it as a pool cue. While I fervently hope that laser pool becomes a sport at some point in the future, radiation pressure is actually used in many applications today, including to propel spacecraft with 'solar sails'. The radiation pressure the solar sails receive as light from the Sun impacts them is akin to the pressure the sail on a boat feels from the wind. This isn't science fiction; this was first demonstrated by JAXA (the Japanese space agency) on their IKAROS (Interplanetary Kite-craft Accelerated by Radiation Of the Sun) spacecraft in 2010. It deployed a 192 m^2 (just over 2,000 square feet) plastic membrane, pointed it at the Sun and managed to fly all the way to Venus.[93] It's an exciting prospect, because there are no moving parts and no fuel that can run out, so craft powered this way could operate for much longer than we're used to.

Radiation pressure is also something space agencies have to take into account when planning missions in the Solar System. Even with a more typically powered spacecraft, say for example on a trip to Mars, radiation pressure from the Sun's light will push it off course, causing the craft to miss Mars by a few thousand kilometres. When spacecraft are launched and set on their merry way, they're sent in slightly

---

93 JAXA reported that the force felt by IKAROS's solar sail was 1.12 millinewtons – equivalent to the same force a pinch of salt feels from the Earth's gravity. This constant force from radiation pressure means the craft is constantly accelerating and accumulating speed. After six months with its solar sail deployed, IKAROS had increased its speed by 100 metres per second (about 360 kilometres per hour) to a top speed of 1,440 kilometres per hour by the time it arrived at Venus. For comparison, the rocket-fuelled Parker Solar Probe arrived at Venus in less than two months after launch and reached it with a speed of approximately 60,000 kilometres per hour.

the wrong direction, knowing light from the Sun will bring them onto the right path.

So the forces from radiation pressure are definitely not something you can ignore. They're enough to power spacecraft and, inside stars themselves during nuclear fusion, enough to resist the crush of gravity inwards. This perfect balance between gravity inwards and radiation pressure outwards in a star is known as the Eddington limit. It's the maximum brightness a star can achieve. If exceeded, the force outwards will be larger than the force of gravity inwards and the star will start to shed some of its outer layers in a wind, or outflow. Because the only thing that radiation pressure in a star needs to resist is gravity, the Eddington luminosity is directly related to the mass of the star. The more massive the star, the brighter it can be.

Similarly, radiation pressure is also a big player in accretion disks around black holes. As the material falls into orbit around the black hole it is accelerated by gravity, heating it up, giving it a huge amount of energy so that it starts to radiate light. This light then exerts pressure outwards on other material trying to fall inwards onto the accretion disk. In a perfect scenario, there'd be a balance between the amount of matter falling onto the accretion disk and the radiation pressure outwards from the material already in the disk. In that case, the black hole would be growing by its maximum possible amount: its Eddington limit. If a glut of extra material falls onto the accretion disk, it will be blown away by radiation pressure, again in a wind or outflow. Black holes therefore have a natural control process to curb their gluttony when their eyes get bigger than their bellies: radiation pressure allows the accretion disk to let a burp rip every now and then.

Just like for stars, the Eddington limit for black holes is set by their mass. The bigger the black hole, the brighter the accretion disk can get, and the faster the black hole can grow (they have a higher 'accretion rate'). A typical supermassive black hole of 700 million times the mass of the Sun would have an Eddington limit (or maximum brightness of its accretion disk) 26 *trillion* times brighter than the Sun.[94] If we assume about 10 per cent of the gravitational energy gained by the matter falling onto the accretion disk is radiated away, then (using $E = mc^2$) we can work out that the maximum rate that a black hole 700 million times the mass of the Sun can grow by is three Suns' worth of material every year.

But that's just a maximum. Only approximately 10 per cent of galaxies have active supermassive black holes at their centre that are currently growing, i.e. they have an accretion disk. And the majority of those are accreting at less than 10 per cent of the maximum rate. Take our own supermassive black hole at the centre of the Milky Way; it is (thankfully) not that active right now. It is radiating at 10 million times less than its Eddington limit, only around a few hundred times brighter than the Sun, meaning it's only growing by a ten billionth of the mass of the Sun every year. A meagre amount.

If there was enough gas funnelled to the centre of the

---

94 This is why the X-rays from accretion disks around supermassive black holes were spotted *way* before any visible light from the billions of stars in the galaxies around them. Trillions >>> billions.

Milky Way, towards the black hole, it could technically grow at a rate 10 million times more than that. But it doesn't: because black holes aren't endless hoovers. *They don't suck.* There has to be some process that physically moves material towards the centre before it gets close enough to be caught up in the accretion disk and brought into orbit by the black hole's gravity. If you think about it, black holes are less like hoovers and more like couch cushions: sat there in your lounge, unassuming, not sucking anything towards them. But if you happen to physically move something close to the edge of that couch cushion and it falls down the back, it's lost down there for good.

12

The old galaxy can't come to the phone right now. Why? Because she's dead

Radiation pressure is a bitch. Not only does it prevent black holes from achieving their full potential but the repercussions can also have a huge impact on the surrounding galaxies. The burps of material let rip by accretion disks around supermassive black holes can be incredibly energetic; enough to shoot out huge radio-emitting jets into intergalactic space that are longer than the galaxy is wide. One such burp that was found by astronomers in March 2020 was the biggest outburst ever seen. It blew out a cavity seventeen times bigger than the Milky Way in the gas between galaxies in a cluster. That'd be like if a human burped in the UK and blew out a cavity in the Earth's atmosphere that went all the way from Newfoundland to the Middle East!

The fact that something so tiny can have such a huge impact is mind-boggling. Let's talk size first: the Milky Way is 100,000 light years across, whereas its black hole is only 0.002 light years across. To put that into context, there's a similar size ratio between a football and the entire Earth. Imagine if the kick of a football could impact the entire planet; that's what we're talking about here with a black hole affecting a galaxy. Sure, the black hole is also supermassive, but compared to the total mass of the galaxy it's a drop in the ocean. The Milky Way's total mass in stars is estimated to be somewhere around 64 billion times the mass of the Sun, but its central supermassive black hole is only 4 million times the mass of the Sun: just 0.006 per cent of the galaxy's mass in stars. And that's just the mass in stars; the total mass, taking into account all the things we can't see like gas, planets, smaller black holes and dark matter, brings the Milky Way's total mass to 1.5 trillion times the mass of the Sun, with the supermassive black hole just 0.0002 per cent of that.

This is why, if you were somehow able to remove the supermassive black hole from the centre of the galaxy, the galaxy wouldn't fall apart. Considering that all the stars in the galaxy are orbiting around the black hole in the centre, that's a little difficult to wrap your head around. If you removed the Sun from the centre of the Solar System then all hell would break loose; but that's because, as we heard in the previous chapter, the Sun is 99.8 per cent of the mass in the Solar System. Lose it and there's nothing holding the planets in orbit anymore and the whole thing would slump apart. But remove the supermassive black hole from the centre of the galaxy and there's enough mass in the rest of the galaxy to hold everything together (something known as self-gravity).

Despite this, the supermassive black hole and galaxy are intrinsically linked: the ratio of their two masses is consistent across the Universe. This was first noticed in 1995 by American astronomers John Kormendy and Douglas Richstone. After collating observations of eight nearby galaxies with active supermassive black holes (including the likes of Andromeda and Messier 87), they noticed a correlation between the mass of the supermassive black hole and the mass of a galaxy's central bulge of stars (you can think of galaxies like a fried egg: they have a beautiful spiral flat disk shape akin to the egg white, and a central blob of stars like the egg yolk). On average, the black holes were a 1,000 times less massive than their galaxies.

Now, eight galaxies aren't exactly representative of the entire galaxy population of the Universe, which is likely in the trillions of galaxies,[95] and so there was a push to measure the supermassive black hole and bulge masses in yet more galaxies to confirm if this correlation was real. This requires being able to work out the Doppler shift from the light emitted by the accretion disk to get at the supermassive black hole mass, and then model how the light is distributed in a galaxy to get at the bulge mass. From how much light you can see, you can then assume a 'mass-to-light' ratio, i.e. if there's this much light then how many stars must there be producing it? To do this, you also need to know what the typical distribution of stars of different masses is in a galaxy (how many massive stars vs. how many smaller stars, on average). It's not a simple task getting all these measurements, but by 1998, thirty-two more galaxies had a bulge mass estimate. This was thanks to the work of Northern Irish astrophysicist John Magorrian, who at the time was working with a giant of the field, Canadian astrophysicist Scott Tremaine, at the

---

95 Although there is a very long-running joke in the astronomy community that three data points make a line. It stems from the scarcity of observations available to people historically.

University of Toronto.[96] Magorrian is now an Associate Professor of Theoretical Astrophysics at the University of Oxford.[97] They used observations from the recently launched (and fixed) Hubble Space Telescope to show that there was indeed a correlation (and a fairly tight one too, as astrophysics goes), with the super-massive black holes[98] around 166 times the mass of the bulges (the Milky Way is actually an outlier from this relationship, with a much smaller black hole than you'd predict for its size).

This correlation is now known as the 'Magorrian relation' and is akin to finding a fossil and learning something new about how life has evolved on Earth. The correlation shows how galaxies and black holes have evolved and grown over the 13.8 billion years of the Universe. It all comes back to a galaxy's bulge; the egg yolk at the centre. Once the initial chaos of formation has settled down, most galaxies will start life as a flat disk of stars all orbiting on nice ordered orbits in the same direction and in the same plane. But if two galaxies get drawn together due to gravity they can merge together, doubling their mass, but disrupting those orbits and the beautiful spiral shape in the process. Through many gravitational interactions, some stars lose energy and sink towards the

---

96 Every astrophysics researcher who studies galaxies will have a copy of James Binney and Scott Tremaine's *Galactic Dynamics*. It's a bible of sorts for us. Arguments are settled with a quick: 'What do Binney and Tremaine say?'

97 I'm beginning to realise just how odd it is to write a book about your own colleagues.

98 Although it's interesting to note that even as recently as 1998, Magorrian referred to these as Massive Dark Objects: MDOs. It's a sobering reminder that my field of astrophysics is still in its infancy.

centre of the galaxy where they form a denser bulge of stars, with haphazard orbits in different directions and planes that resemble a swarm of bees.

The two supermassive black holes also merge as the galaxies merge,[99] growing in mass. But just as the stars interact to sink them to the centre, so do gas particles, which find themselves funnelled into the black hole's accretion disk so that it can grow further. It's this joint growth of the galaxy and its black hole in a galaxy merger that's thought to be responsible for the Magorrian correlation between the two. This idea is known as 'co-evolution' of galaxies and black holes. My own recent work has challenged the idea that mergers are the only process that can drive this co-evolution. Along with my colleagues Brooke Simmons and Chris Lintott, we observed some galaxies with no bulge, and therefore no merger, and showed that they have supermassive black holes as massive as those that have had a merger. We then collaborated with some theorist friends[100] who simulated this non-merger growth and found that it could explain 65 per cent of all supermassive black hole growth in the Universe. It's likely that mergers aren't the dominant force driving this correlation between black

---

99 The probability of any two stars physically colliding in a galaxy merger is vanishingly small, because once again, space is just very, very big.

100 In the Horizon-AGN simulation team, including Garreth Martin, Sugata Kaviraj, Julien Devriendt, Marta Volonteri, Yohan Dubois, Christophe Pichon and Ricarda Beckmann. Ricarda and I also did our PhDs together in Oxford; we were roommates for two years and now collaborate together on our research, as well as remaining good friends.

holes and their galaxies, but leave it with me a bit longer, to work out what process is responsible instead![101]

Regardless of what's driving it, this correlation has now been seen across a huge population of galaxies thanks to observations from huge astronomical surveys. These are telescopes that are not at the beck and call of astronomers around the world to observe a few objects as part of their current niche project[102], but telescopes that observe the entire sky night after night, slowly building up a mosaic image of the entire sky, detecting fainter and fainter objects with every pass. This allows you to build huge catalogues of the positions, images and spectra of all the stars and galaxies visible from that part of the world. One of the largest of these surveys (and one of the largest collaborations of astronomers in the world) is the Sloan Digital Sky Survey[103] (SDSS) which uses the 2.5-metre optical telescope at Apache Point Observatory in the middle of the Sacramento Mountains in New Mexico. In its first data release in 2003, it provided observations of

---

101 Remember, science needs time. And funding; if anyone out there at a university wants to offer me a fellowship or permanent professorship to figure this out? I know, I know, I'm a shameless postdoc.

102 I say beck and call, but putting together a proposal to use a professional telescope is a very lengthy process and there's no guarantee of time when telescopes are horrendously over-subscribed. The VLT in Chile, for example, is over-subscribed by a factor of eight on average in each round of proposals.

103 Named after the Alfred P. Sloan Foundation, which was set up in 1934 by Alfred P. Sloan Jr., who was then the president and chief executive officer of General Motors. The foundation awards grants to projects across science, technology and engineering disciplines.

134,000 galaxies across the northern sky, including over 18,000 quasars. By 2009, those numbers had jumped to just under one million galaxies and over 100,000 quasars.

The realms of large-number statistics had been opened to astronomers by surveys like SDSS, and they allow us to study populations of growing black holes to understand what their effect on galaxies truly is. Observations with SDSS confirmed the Magorrian relation, but also showed that the mass of the supermassive black hole also correlates with the total mass in stars in a galaxy, not just in its central regions. One thing these big surveys noticed, though, is that there is a steep drop-off in the number of the most massive galaxies. These are the galaxies that are 100 per cent bulge. They've had so many mergers that their spiral shape has been completely destroyed and what's been left is just one giant big blob of a galaxy.[104]

This distribution of the different masses of galaxies is called the 'luminosity function' (since mass is intrinsically tied with how bright a galaxy is and it is brightness that we measure directly), and to figure out what shape it is you have to first know how galaxies form and at what masses, and how they evolve after that. The people who originally tried to have a crack at predicting what gave us these differing numbers of smaller and more massive galaxies were British astrophysicists Martin Rees and Simon White, along with American Jerry Ostriker[105] in the late 1970s. That trio is a dream dinner party

104 The technical term is 'elliptical', but I prefer blob. Especially because I hear it in Rowan Atkinson's voice in my head.
105 Jerry Ostriker is the husband of celebrated American poet Alicia Ostriker, known for her Jewish feminist poems.

right there. Rees is the current serving Astronomer Royal, and previously served as Master of Trinity College, Cambridge, and President of the Royal Society. White was a PhD student at Cambridge at the time and has since served as one of the directors of the Max Planck Institute in Garching, Germany. Ostriker completed his PhD at the University of Chicago in the late 1960s with none other than Subrahmanyan Chandrasekhar (of the maximum mass of a white dwarf fame) and has served as professor of astrophysics at Cambridge, Princeton and Columbia, along with a stint as provost of Princeton University. They definitely qualify as BNIPs. Together they came up with a model for how galaxies form in the early Universe as gas clouds start to cool down; if a gas is too hot it can resist the pull of gravity down and it won't become dense enough to form stars.

Rees, Ostriker and White thought that the cut-off in the luminosity function at high masses could be explained if the most massive galaxies formed from the most massive gas clouds. They reasoned that the Universe hasn't been around long enough for these most massive gas clouds to have had enough time to cool yet. Their basic model of cooling gas clouds would be continuously fine-tuned over the next few decades by a host of astrophysicists, to encompass mergers of gas clouds and the effect of newly formed stars (which put out more heat to stop the gas clouds from cooling). By the early 2000s, astronomers had a realistic model and, crucially, enough computing power to simulate galaxies forming and evolving in the Universe.

You can then directly compare a computer-simulated

universe to the observed Universe to check if you got everything right; including the shape of the luminosity function by just counting how many galaxies of each mass you form. It quickly became clear that the two did not match in the slightest. There were far too many high-mass galaxies in the simulated luminosity function. That meant the simulations were missing something; either some law of physics coded into the simulation was wrong or a process affecting galaxies was unaccounted for.

At the forefront of the development of these simulations was a group of astrophysicists at Durham University's Institute for Computational Cosmology, including Carlos Frenk, Cedric Lacey, Carlton Baugh, Shaun Cole, Richard Bower and Andrew Benson.[106] Together, they realised that the missing process in the simulations was the energy injected by outflows driven by radiation pressure from the accretion disks around supermassive black holes. By 2003, they had managed to incorporate this into their simulations and show how they recreated the steep drop-off of the luminosity function: their simulation no longer over-produced massive galaxies.

The idea goes that the outflow of radiation and material from accretion onto the black hole can either heat up the gas (so that it can't cool and collapse down to make new stars), or eject it from the galaxy entirely. Either way, the effect

---

106 Frenk, Lacey, Baugh and Cole all taught me an area of physics while I was an undergraduate at Durham University. That's one of the wonderful things about being a student – getting taught by experts who are at the cutting edge of research. Not that you're fully aware of it at the time.

would be to quickly shut off star formation in a galaxy, at least in galaxies with the most massive supermassive black holes which, as we've heard, are in the most massive galaxies. We call this a 'feedback' effect, because as the galaxy feeds the black hole, the black hole can in turn throw energy back out that has a negative effect on the galaxy; the galaxy essentially shoots itself in the foot. It's this feedback that's thought to regulate the co-evolution of both galaxies and their central black holes, stopping both from getting too big for their boots.

With many other simulation groups managing to recreate the Durham group's result, this feedback hypothesis became accepted among the theoretical astrophysical community. The problem is that among us observational astrophysicists that use telescopes to take data on the real Universe, we haven't found any evidence for this happening. There have been some individual cases in single galaxies where the negative effects of an outflow or jet from an accretion disk can be seen (sometimes even causing shocks that compress gas so that new stars form in a process called positive feedback), but not in the huge population-wide studies, for example using data from a big sky survey like SDSS, that would allow us to make conclusions about the Universe as a whole. I should know; this is exactly what I spend the other half of my research time on, trying to find statistical evidence for negative feedback, helping to add my own little nuggets of insight into the collective astrophysical knowledge like all those before me.

One way we can tell whether an outflow from an active supermassive black hole has had an impact on a galaxy is to look at its colour. As we learnt right at the beginning of this

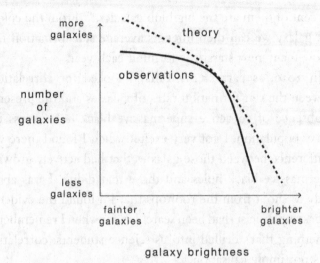

A cartoon 'luminosity function' showing the number of galaxies found at each brightness that we observe in the Universe (solid line), compared to the number originally found in simulations (dashed line). Initially, simulations over-predicted the amount of very bright and very faint galaxies, revealing that there was some physical processes that had been missed.

book, the most massive blue stars live much shorter lives than the smaller red stars. So if you look at the colour of a galaxy overall and it looks very blue, you know it must have formed new stars recently. Whereas if a galaxy appears red overall then you know that enough time must have passed that the more massive stars have died and gone supernova, leaving just the smaller, longer lived, redder stars; like the dying embers of a fire. We refer to galaxies that have stopped forming stars as 'red and dead', and interestingly we find that around 70 per cent of them are the big blob galaxies.[107] From the colour of a galaxy we can then infer its average star formation rate – how many new stars it is forming each year.

In 2016, as part of my PhD, I looked for correlations between the star formation rates of galaxies and the presence or absence of an active supermassive black hole across the galaxy population. I got very excited when I found there was a difference between those galaxies that had actively growing supermassive black holes and those that didn't. I was about ready to shout from the rooftops that I'd found the evidence that astrophysicists had been searching for when I remembered something that's drilled into us science students: correlation does not imply causation.

For example, the sales of ice creams and sunglasses are correlated. Do you put on your sunglasses and immediately want an ice cream because of it? Or eat an ice cream and then wish to look as cool as your frozen dessert? No. The two are correlated because they're both caused by the fact that the weather is warm and sunny. Remembering this fact, I realised that what I'd found was evidence for a shutdown of star formation at the same time as the black hole was active – what if another process was what had actually caused both things? Something that had managed to both heat the gas to stop stars forming and at the same time funnel gas towards

---

107 For most of the twentieth century, all red galaxies were thought to be blob shaped. It was the work of British astrophysicist Karen Masters and the Galaxy Zoo team using images from the Sloan Digital Sky Survey that showed that around 30 per cent of red galaxies are actually spiral shaped; you don't need a merger to shut off your star formation.

the centre to feed the black hole. Maybe a merger of two galaxies? Or something else entirely again?

So now in my research, I'm trying to find the smoking gun of supermassive black hole feedback, something that's irrefutably caused by the outflow itself. To do this, I've joined a worldwide collaboration of astrophysicists working with the telescope that conducted the Sloan Digital Sky Survey. It recently finished a new survey, called MaNGA,[108] where instead of taking one single observation of a whole galaxy, it takes over a hundred, mosaic-ing a galaxy to observe each region individually in over 10,000 galaxies. No longer are we resigned to reducing a complex system of billions of stars to a single measurement; we can peer into the inner workings of a galaxy to answer the unanswered questions still plaguing those studying the evolution of galaxies.

My niche area of that collaboration is trying to trace the

---

108 The brainchild of American astrophysicist Kevin Bundy, now an Assistant Professor at UC Santa Cruz, who is a brilliant champion of all us working in the MaNGA collaboration. I first met Kevin at a conference on the Mexican island of Cozumel that was being held at an all-inclusive resort. There was a pool with a swim-up bar at this hotel, and us PhD students (as I was at the time), desperate to make a good impression with the more senior academics, diligently ignored the bar and attended all the research sessions instead. However, we quickly realised that the best way to network at this conference was not to attend the sessions, but to show up at the bar, because that was where all the senior academics had obviously drifted off to. I remember grabbing a Piña Colada, swimming up to a group of people chatting and introducing myself to the nearest person: 'Hi, I'm Becky!' – 'Hello, I'm Kevin Bundy' – I nearly choked on my drink.

feedback effects that supermassive black holes have on galaxies. Is there a correlation between the star formation rate in a given area and the distance from the black hole in the centre? Does that drop in star formation trace the energy of the outflow as it moves through the galaxy? If this effect is there, is it more appreciable in galaxies with more massive black holes? These are the unanswered questions I spend my days pondering over. It's complicated stuff and easy to get frustrated with a lack of progress. But breakthroughs don't happen overnight; the history laid out in this book is a testament to how slow and steady wins the collective human knowledge race.

With time, my colleagues and I will all analyse our data and publish our results, which will collectively come together to give us the big jigsaw puzzle picture of what is going on. Are the outflows from accreting black holes responsible for shutting off star formation in their galaxies to make them 'red and dead'? Either way, something has been killing galaxies: *there's been a murder*. And us astrophysics detectives will crack the case.

13

You can't stop tomorrow coming

Everyone has a favourite word. That combination of syllables, consonants and mouth shapes that somehow sparks joy. Tolkien might have had his precious 'cellar door' combination, but for me, my favourite word in the entirety of the English language is 'spaghettification'. My mouth has to work overtime to even be able to say it, my fingers have to blaze across the keyboard to type it and my brain has to think really hard to remember how to spell it.[109] But I dare you to say it without breaking into a smile. You may even find yourself channelling Sean Connery as you proclaim 'spaghettification!'

Now, as much as this sounds like a word I have made up to make myself happy, it is in fact a real astrophysical term; a phenomenon that black holes cause. All this information about black holes may have got you as excited as I am about them. You may even be wondering what it'd be like to visit a black hole, or perhaps get close enough to peer beyond the event horizon. Well, let me warn you now reader, that that is something you would never wish to do, for fear of being spaghettified.

The gravity around a black hole is so strong that if you

---

109 Space is hard, words are harder.

fell towards it head first the gravity would be so much stronger at your head than at your feet that you would get stretched out like Elastigirl from *The Incredibles*. You would look more like spaghetti than a human; a long thin chain of atoms stretching all the way down to the centre of the black hole. We've seen this happening to gas clouds like G2 around the Milky Way's central black hole, but also to stars, as they go from perfectly spherical to stretched thin.

This is all because of the gradient of the strength of the gravity around a black hole. Far enough away and it's no different than the pull of a planet or a star, but get too close and the increase is exponential. It's this gradient that causes the spaghettification; imagine being at the top of a very steep water slide, holding on where it's flatter, but your feet are all the way at the bottom, lost over the edge. Strangely, with spaghettification it's the less massive black holes you have to be more wary of, rather than the supermassive ones.

As a black hole gets more massive, its event horizon gets larger. The area of space that the black hole influences is much bigger, but the gravity gradient doesn't get really steep until very near the black hole, sometimes well within the event horizon itself. But a less massive black hole has a smaller event horizon, and the gradient of gravity can get really steep outside it. Gravity is not stronger around the smaller black hole, but the strength changes more rapidly with every step closer you take. Think of it like mountains; the height of one mountain can be less than another, but the climb up can still be much steeper.

Or for the skiers out there, getting closer to a less massive black hole would be like cross-country skiing on the flat for a while before the slope all of a sudden becomes a super-steep black diamond run that could injure you. Thankfully, though, there's a ski lift to take you away from danger (because in this analogy you haven't crossed the event horizon yet). But getting closer to a supermassive black hole would be like being on a gentle beginner's slope for a long time, before it gradually transitions to a steeper blue slope, then a steeper red slope, and then finally a super-steep black run that you could hurt yourself on but you realise too late there's no ski lift to take you away and the only way is down. The Milky Way's super-massive black hole is on the piddly side of supermassive, hence why the G2 gas cloud, when it looped around it back in 2014, got a bit spaghettified but otherwise escaped unscathed (it got the ski lift out of there, to labour that analogy).

So if you did desperately want to feel the effects of spa-ghettification you could in theory get closer to a less massive black hole and still escape, but your very shape would be irrevocably changed. This is what you would feel if you 'fell into' a black hole, but what would you actually see? Assuming you could somehow resist the stretching effects, perhaps in a spaghettification-proof spacecraft,[110] what would you see out of the window? Well, thanks to general relativity, we have the equations to work out what would happen without any astronauts needing to make the ultimate sacrifice.

Let's assume the black hole we're falling into is not accreting material, so that we don't blind/kill ourselves with

---

110 Patent pending.

high-energy radiation sat in the window seat of our spacecraft. From a great distance you wouldn't see much, black holes are very dense after all, so size-wise they're pretty small and you wouldn't be able to spot them from far out. As you got closer, though, you would eventually be able to notice a small dark circle where there was no light whatsoever, marking the event horizon.

As you got ever closer to the black hole you'd start to think your mind was playing tricks on you. Black holes curve spacetime so much that they affect the path of light from behind and around them, messing with your sense of perspective. Approaching a typical object in space, like the Moon for example, on your journey from Earth, the object would get steadily bigger in your window directly proportional to how close you were. When you were halfway there, the Moon would look twice as big as it does on Earth. But with all the curving of light around them, black holes don't behave the same as the Moon.

Black holes are like puffer fish; they make themselves look bigger than they truly are. The light from stars behind them gets bent to the side so that the area where no light is coming from looks bigger than it really is; an effect that is exacerbated as you get ever closer. So much so that if you were ten times the event horizon away from a black hole, the black hole would completely block your view looking out of your spacecraft window. Compare that to being ten times the size of the Moon away from the Moon, and the Moon would be about the size of your fist held in front of you at arm's length.

Getting closer still, the black hole would continue to appear larger around you, with the darkness slowly engulfing your spacecraft from all angles as the black hole continues to bend light out and away from you. Looking backwards, you'd see not only the view back the way you came, but the view behind the black hole, bent into your eye line. A 360° view squished into an ever-shrinking circle, until at the event horizon it becomes a single dot of light: the light of the entire Universe bent into your eyes for one final glimpse, one look back over the shoulder, before you face the unknown.

I can't tell you what happens next, as you cross the event horizon. Do you descend into darkness or into bright blinding light? Is there a star-like object there made of an exotic form of matter that we don't know about that's held up by another form of degeneracy pressure: the next stage in a star's evolution from white dwarf to neutron star to *something else*? Has all the matter that has been trapped beyond the event horizon for billions of years turned to pure energy? Is there really a singularity? Only you would know, having crossed over, but you'd never be able to share what you found with the world.

Once beyond the event horizon, every direction would be 'downhill'. Even if you turned around the way you came, every path would lead you to the centre. Perhaps you might panic and try to accelerate away from the centre to get back out, but that would just get you to the centre even faster. There's no way out. Every single version of your future has you ending up at the very centre of the black hole. Space and time become one so that the future is a direction in space rather than in time. Your spacecraft wouldn't be able to save you, just like it wouldn't be able to stop tomorrow from coming.

That's your perspective though: what you would see. But what if you had a friend who wanted to watch from a safe distance what happens when you fall in? Perhaps you might set up a system where you send a burst of light every minute, like a lighthouse, just to let them know you're OK on the journey. To you in your spacecraft, you will send off those bursts every minute, on the minute. But that's not what your friend will see. Because to you, getting ever closer to the black hole and its strong gravity, time would pass differently than to your friend at a safer distance. What feels like a minute to you could be an hour or more from their perspective.

This is something called time dilation; a concept that Einstein explained for moving objects in his theory of special relativity way back in 1905. This phenomenon had already been predicted for electrons orbiting atoms in 1897 by Northern Irish physicist Sir Joseph Larmor, but it was Einstein that linked this back to the very nature of time itself, as opposed to a property of electrons. Einstein derived the relation between the difference in time that passes and the difference in speed that two objects are moving at. The greater the difference in speed, the bigger the time difference, so much so that as you reach the speed of light time slows to a standstill.

The speeds that we can currently achieve for space travel don't produce a time dilation that's noticeable by current astronauts. For example, astronauts aboard the International Space Station, which orbits at an average altitude of 408 km at a speed of 27,500 km/h (17,000 mph) will experience around 0.01 seconds less time than those on Earth for every year they spend in space. After a year onboard, they touchdown back on Earth 0.01 seconds younger than they would be if they'd stayed at home.

This is called 'kinetic time dilation', an effect caused by increased speed. But there is a second type of time dilation: 'gravitational time dilation'. Instead of a higher speed causing time dilation, it can also be caused by incredibly strong gravity; the stronger the gravity, the slower time passes for you relative to someone in lower gravity. This effect is not just noticeable around black holes; the gravity at the core of the Earth is stronger than at the crust, making the core ever so slightly younger than the crust. It also means that astronauts on board the International Space Station in a lower gravity than us on the ground experience time a little faster, actually cancelling out the effects of kinetic time dilation making them younger.

Time dilation has been tested and proved many times in many ways over the past century, but perhaps the most famous experiment was one designed by two Americans: physicist Joseph Hafele and astronomer Richard Keating. Hafele was an assistant professor in St. Louis in 1970 when he was preparing a lecture for students on relativity and time dilation. He ended up doing a quick calculation on the amount of time dilation a commercial airliner would experience with a typical airspeed of 300 m/s (670 mph) at a typical altitude of 10 km (33,000 ft). He realised the combination of time slowing down due to kinetic time dilation and time speeding up for the lower gravity would give an overall time difference of around 100 nanoseconds (0.0000001 seconds; remember human reaction time is 0.25 seconds, so this is a tiny fraction of a second).

To measure such a tiny difference, you need an incredibly precise clock; one that can measure to nanosecond precision. In 1955, the first such clock was built at the National Physical Laboratory in south-west London, using caesium atoms as the inbuilt time keeper. It's not just the light from stars that can make electrons in atoms jump up orbits into excited states; we can use lasers to do this too. The electrons absorb a little bit of energy, jump up an energy level and then drop back down, emitting a very specific wavelength (or colour) of light. This is how we know what elements are present in nebula gas clouds that form stars; specific elements emit specific colours, like a fingerprint.

You can fine-tune this process even more; if the laser you use has the same wavelength of light as the wavelength given off by the electrons as they jump down, you hit a sweet spot and give the electrons just the right amount of energy to keep oscillating between their excited and normal states. We describe this as the atom and the laser being in resonance. If you can find that wavelength sweet spot with your laser then you know the exact frequency that the transition happens at, thanks to the wave speed equation that we all learn at school. For light, the speed of light is constant and so frequency and wavelength are intrinsically linked: speed of light = frequency × wavelength.

So for caesium atoms, we've found the laser wavelength sweet spot and know that the electrons jump up and down between the first two orbits when they're in resonance 9,192,631,770 times a second. This is so precise that although the second used to be defined based on the Earth's rotation as 1 / 86,400 of a single day, the second is now defined by a caesium atomic clock because it's more precise (it's also measurable anywhere in the Universe as well). Today's caesium atomic clocks are so accurate that even in 100 million years, they won't drop or gain a second (compare that to a typical mechanical wrist watch that drops around five seconds a day on average).

Back in 1970, atomic clocks weren't as precise as they are today, but could still measure time to a few nanoseconds of precision. Hafele realised that two out of three things he needed to easily test the time dilation predictions of relativity were readily available to him: airplanes and atomic clocks. The third thing, which wasn't readily available to him, was money. He spent another year as an academic beggar, asking many institutes for money to do the experiment, before meeting astronomer Richard Keating, who worked in the atomic clocks department at the US Naval Observatory. Atomic clocks were also used for nautical navigation at the time, as a much more useful replacement for the timing of Io's eclipses. Keating helped Hafele obtain $8,000 of funding from the Office for Naval Research, $7,000 of which was spent hiring out commercial aircraft and crew. On each flight they had a seat for both Hafele and Keating and two seats for a passenger named 'Mr Clock'.

They flew the atomic clock heading east around the world, and then two weeks later flew around the world heading west, comparing the time recorded on each clock to others that had been kept on the ground by the Naval Observatory. In this experiment, the airplanes are moving and the centre of

Earth is the stationary reference point, since it doesn't move as the Earth rotates. A plane flying east, in the same direction as the Earth rotates, has a higher relative velocity than a plane flying west in the opposite direction to the Earth's rotation. So on the two flights, a different kinetic time dilation should occur (with the clock on the eastern flight losing time compared to the western flight). Combining this with the much stronger effect of gravitational time dilation (assuming the two planes fly at exactly the same constant altitude, which in reality won't quite be the case) gives a total predicted time loss of 40 nanoseconds on the eastern flight and 275 nanoseconds gained on the western flight.

Hafele and Keating published their results in 1972, reporting a measurement of the time lost on the eastern flight of 59 nanoseconds (±10 nanoseconds due to measurement error, meaning the value could be anywhere between 49–69 nanoseconds) and time gained on the western flight of 273 nanoseconds (±7 nanoseconds). The agreement between predicted and measured values in this experiment is astonishing and is one that has been repeated many times since with the same results. It showcases just how accurate the predictions we can make with Einstein's theories of special and general relativity really are. And a good job too, because GPS satellites in orbit around Earth suffer from this same kinetic and gravitational time dilation (the gravitational time dilation is what dominates); the clocks on board the satellites gain 38,640 nanoseconds per day compared to clocks on Earth. If we didn't correct for this time gain, then GPS would be utterly useless in giving an accurate position within two minutes. These errors in positions would snowball by 10 km per day (or around 6 miles).

So even just above our heads, here on Earth relativity has a noticeable effect. Imagine, then, the effect of gravitational time dilation around a black hole a trillion times more massive than Earth. You on your spaghettification-proof spacecraft sending out a flash of light once every minute to your friend watching your journey towards the black hole, would not notice any difference in how time flowed. It would truly still feel like one minute to you and not like time had slowed down. But to your friend, those light flashes would take longer to arrive as your speed appeared to slow down as you got closer to the event horizon. A minute between flashes would turn to an hour, an hour to a day, a day to a year and a year to a century. In fact, someone watching you get ever closer would never actually see you cross the event horizon, appearing as if time had frozen for you, when in reality you crossed with no bother, feeling like only a few hours or days had passed since the start of your journey. The event of your light flash when you cross that point of no return would be for ever outside your friend's possible powers of observation.

This freezing of time is an optical illusion created by the effects of gravitational time dilation, like the illusion of the black hole appearing much larger out of the window due to the curvature of space. Black holes in that sense really are the ultimate tricksters; we can't rely on what we see. Instead, the equations of general relativity can open the door and reveal the truth, no matter how massive the black hole.

14

Well, Judy, you did it.
She's finally full

At four million times heavier than the Sun, the black hole at the centre of the Milky Way might sound impressively large, but it is far from the biggest. The only black hole we have an image of (at time of writing anyway) is the supermassive one at the centre of the M87 galaxy that we saw back in Chapter 10. It's the galaxy at the very centre of a supercluster of galaxies that includes the Milky Way; if you could keep zooming out from Earth to see the big picture, then at the centre of everything there would be M87's supermassive black hole. The age-old phrase 'all roads lead to Rome' should really be 'all roads lead to black holes'.

M87's black hole is 6.5 billion times more massive than the Sun. It makes the Milky Way's black hole look like a lightweight. But even that is still not the biggest. The heavyweight crown goes to TON 618, which is 66 billion times the mass of the Sun. It is so big that astronomers had to invent a new word for it – an *ultramassive* black hole. But as we've already heard, black holes aren't just endless hoovers: they don't suck. They're limited to how fast they can grow because of radiation pressure (the Eddington limit).

We know that the majority of black holes don't accrete at the Eddington limit, at their maximum rate, because of radiation pressure pushing material back out. When we look at the distribution of the growth rate of active supermassive black holes, we find that on average they take on material at about 10 per cent of their maximum possible rate. So can black holes just endlessly grow at that rate with no limit to the mass they can reach? Technically, the theoretical maximum would be a black hole containing all the mass in the entire Universe. That number is a bit difficult to estimate, but it's somewhere around the 10^{60} kilogram mark. That's a 1 with sixty zeros after it; a *novemdecillion* to use its technical term.

I feel it's my civic duty to point out here that a black hole with a mass of novemdecillion kilograms is highly unlikely. Space itself is expanding, taking galaxies, and therefore the matter in the Universe, ever further apart. This will reduce the amount of material available for black holes to eventually accrete; once they've exhausted the supply their galaxy can give them, then that's it. It also reduces the likelihood of any galaxy mergers as the Universe ages, and therefore there'll be fewer supermassive black hole mergers to go with them. A merger can at most double the black hole's mass, so it's a very efficient growth process, but the occurrence is getting rarer with every passing day.

The growth of supermassive black holes is very reliant on accretion; the process of taking in more matter through all those collisions between gas particles in the accretion disk, to slowly reduce their energy and bring them closer to the black hole. If you disrupt that process in any way then that's it for the black hole – it can no longer grow any bigger unless it gets lucky with a merger. So are there any processes that can disrupt this accretion process? And if so, what's the maximum mass of a black hole then?

The first people to try and put an estimate on this maximum mass were Indian astrophysicist Priya Natarajan (now a professor at Yale University) and Argentinian astrophysicist Ezequiel Treister (now a professor at the University of Chile) in 2008. They argued that a limit to a black hole's mass naturally occurs because of the co-evolution of supermassive black holes with their galaxies. With continued growth of the black hole comes continued feedback, which eventually blows away the accretion disk around the black hole. They estimated that this would mean that a black hole could only reach up to 10 billion times the mass of the Sun.

But in 2015, British astrophysicist Andrew King entered the chat. King did his PhD at the University of Cambridge during the heyday of black hole research in the 1970s, working with Stephen Hawking. He's now a professor at the University of Leicester and in 2014 was awarded the coveted Eddington medal from the Royal Astronomical Society for his work on black holes and general relativity. King pointed out a quirk of gravity around a black hole that allowed him to estimate the maximum mass a black hole could grow to via accretion as 50 billion times the mass of the Sun (but that could be pushed to a whopping 270 billion times the mass of the Sun if the black hole was spinning in the same direction as its galaxy).

It's all to do with the many different 'spheres' you can draw around a black hole. We've heard about the event horizon already; what we define as the size of the black hole because it's that point of no return where we no longer receive any light. But there's a few more distances from the singularity that get thrown about in casual astrophysical conversations. There's the ergosphere – the region around a black hole you can extract energy from (perhaps obvious to those who speak Greek, *ergon* means work), for example through a gravitational slingshot like spacecraft use in our own Solar System to steal away a bit of energy from something much more massive than them.

Then there's the photon sphere – the region around the black hole where gravity is so strong that any photons (particles of light) travelling at the speed of light would have their path curved so much that they'd travel in a perfect circle. Theoretically it would be possible to see the back of your own head at the photon sphere (if you hadn't been spaghettified[111] first). This sphere is just beyond the event horizon, about 1.5 times larger.

But crucial to the process of accretion is the sphere called the Innermost Stable Circular Orbit, or ISCO.[112] In Newton's version of gravity that we all learn at school, all perfectly circular orbits, no matter the distance, are very stable. That means if an object on a circular orbit is perturbed slightly, let's imagine a rather large asteroid impacts with another asteroid on a perfectly circular orbit, then the orbit can adapt and become slightly elliptical (remember, a circle is just a very

111 Any excuse to drop it into conversation.
112 It almost sounds like a beatbox percussion. Lin-Manuel (I've mentioned him so many times in these footnotes we're on first name terms now), I am patiently waiting for a black hole hip-hop musical that can make possible a beatbox number all about the ISCO.

special case of an ellipse where the aphelion equals the peri-helion). So, that would mean that even if something was somehow orbiting the Sun in a perfect circle just above its surface, and was nudged somehow, it could still adapt its orbit to an elliptical shape to continue orbiting the Sun.

In Einstein's general relativity, though, that's not the case. As you get closer to an object, and in particular a very compact object like a black hole, there is a point where if you nudge something on a circular orbit, it can't correct and it ends up spiralling inwards to the black hole. This is the ISCO, and it sits at three times larger than the event horizon (although if the black hole is spinning that can shrink slightly). Anything that has mass (i.e. not photons of light) cannot form a stable orbit around a black hole any closer in than the ISCO. Usually this marks the rough edge of the accretion disk around the black hole. Just like with the event horizon, the ISCO is related to how massive the black hole is. As the black hole grows in mass, the ISCO gets pushed further out.

There's one more circle around a black hole to mention: the self-gravitational radius. Now, this also depends on the object that's creeping too close, along with the mass of the black hole, but essentially it marks the point at which the pull of gravity holding the object together (self-gravity) is stronger than the pull from the black hole. This is a really crucial point because it explains why we even have galaxies of stars surrounding supermassive black holes in the first place; beyond this radius gas in a galaxy is attracted to itself, more than to the supermassive black hole in the centre, and so the gas can get denser before collapsing in on itself to form stars. If this wasn't the case then we wouldn't be here at all, our atoms would all just be part of one giant accretion disk around the supermassive black hole of the Milky Way.

What Andrew King pointed out in 2015 was that, as supermassive black holes grow ever bigger (via accretion and co-evolution with their galaxies), the ISCO gets pushed beyond the self-gravitational radius. What that means is that any gas particles in the accretion disk, no matter how many collisions they have, will never lose enough energy to reduce their orbit enough that they will reach the ISCO and spiral inwards to grow the mass of the black hole. Instead, the pull of gravity from all the other particles in the accretion disk will always be stronger than the pull of gravity from the black hole.

In fact, at this point, an accretion disk isn't even going to form. Instead, if you have an influx of gas, its self-gravity will hold it together and it will loop around the black hole relatively unscathed; akin to the G2 gas cloud's trajectory around the Milky Way's black hole. Unless material is on a direct trajectory with the black hole at the bullseye (which is rare given how big space is and how relatively small black holes are, even the ultramassive ones) it won't become part of the black hole. This lack of accretion disk means that we also won't be able to spot an ultramassive black hole, as there'll be no luminous matter around it lighting up like a Christmas tree.

This is what makes TON 618 so interesting; with an ultramassive black hole of 66 billion times the mass of the Sun, it lies above King's estimate of the maximum limit for a non-spinning black hole (which was 50 billion times the mass of the Sun). Most black holes are spinning (angular momentum, you just can't shake it), so that's not unsurprising, but it does mean it could be nearing its maximum mass.

TON 618's peculiarity was noted well before it was recognised for what it was. It was spotted on photographic plates taken in 1957 at the Tonantzintla Observatory in Mexico by Mexican astronomers Braulio Iriarte and Enrique Chavira, who noted that it looked violet in colour. It was finally identified as a quasar in 1970 by a group of Italian astronomers conducting a radio survey of the sky in Bologna. By 1976, French astronomer Marie-Helene Ulrich had managed to use the McDonald Observatory in Texas to calculate its distance (the light left it 10.8 billion years ago) and work out it was one of the most luminous quasars ever known (the more luminous the quasar, i.e. the accretion disk, the more massive the black hole).

It's by using the measured speed of the gas in the accretion disk that the estimate of TON 618's mass is derived: 66 billion times the mass of the Sun. I know I keep repeating that number but it really is *huge*. It's more than the total mass of stars in the entire Milky Way (estimated at 64 billion times the mass of the Sun). Its event horizon is 1,300 times larger than the Earth–Sun distance (forty times the distance of Neptune from the Sun). It's a behemoth: massive enough to spark fear into the hearts of us puny humans, and yet unless you launched yourself out of a canon, Zazel-style,[113] directly

---

113 Rossa Matilda Richter, also known by her stage name Zazel, was the very first person to be shot out of a cannon at the age of seventeen in 1877 at the Royal Aquarium, London. She toured Europe and America with Barnum & Bailey's travelling circus, aka 'The Greatest Show on

into TON 618, there's absolutely nothing to fear from it. Almost as if the Universe finally put a stopper in the sink plug hole.

It's fascinating to consider the implications of this maximum mass a black hole can grow to via accretion, and that TON 618 has even come near to it. It means we could be approaching the epoch of the Universe, where black holes reach their limit. As black holes reach this limit and quit growing, or glowing, quasars across the Universe will begin to wink out. If this had happened just a few million years earlier, we as humans may never have even known supermassive black holes existed. It could even be the case that there are some black holes that have reached ultramassive status, but we don't know they're there. Without some sort of light from the accretion disk, we cannot hope to measure the mass of the black holes in the centres of distant galaxies. Perhaps ultramassive black holes are already hiding among us.

I am both equally amazed and at the same time slightly disappointed that we are right now living through the epoch of the Universe, where some black holes might never grow any bigger. It's as if these big, scary, mysterious, infuriatingly interesting black holes are past their heyday, over the hill, senescent. I don't know whether to laugh or to cry at the thought. And yet they might just have the last laugh.

Earth'. Fans of the recent Hugh Jackman film *The Greatest Showman* should be familiar.

15

Everything that dies, someday comes back

E ternity is a very long time. The human brain can't really wrap its head around the concept of infinity. Especially infinite time; no matter how many novels are written contemplating the idea of immortality. When thinking about how black holes form and grow, it's inevitable that we wonder whether they too can die. Are black holes eternal and immortal, living for ever as the Universe evolves, with matter forever trapped inside the prison of the event horizon? Or is there a way they can eventually die?

It was British physicist Stephen Hawking who contemplated this question in 1974. Hawking's life is truly extraordinary. In 1963, at the age of twenty-one, just six months after starting his PhD in cosmology at Cambridge University, Hawking was diagnosed with early onset motor neurone disease; an affliction that limits someone's control of their voluntary muscles, which control speaking, eating and walking. His doctors advised him he had two years to live, and at that point he felt there was little reason to continue with his studies. With his disease progressing slower than first thought, and his mind unaffected, his PhD supervisor Dennis Sciama encouraged him to return to his research on singularities. In his PhD thesis, Hawking explored the idea that the

Universe itself could have started in a singularity, an idea that would revolutionise cosmology through the application of general relativity.

With the discovery of neutron stars in the late 1960s and with Hawking's work on singularities (applied to both black holes and the beginning of the Universe), the idea of a black hole was becoming more generally accepted, at least in the theoretical physics community, but there were many questions. Black holes, when you first consider them, seem to break many laws of physics, one of the most basic being the second law of thermodynamics: that entropy must always increase. Entropy is often described as a measure of disorder, but a better description might just be that whatever is the most likely thing to happen, will happen. If you fill a box with coins all heads up and shake it up, it's very unlikely all the coins will stay heads up, or end up all tails. Instead, the most likely thing that will happen is that you'll end up with a mess of roughly half heads up and half heads down. Crack an egg into a jar and then shake it up, the most likely thing that will happen is that the yolk will not remain whole. This is an especially good example, as the process of scrambling the egg is irreversible: you can't unscramble the egg because entropy cannot decrease.

As matter is accreted by a black hole, it is locked away nice and neatly beyond the event horizon for evermore. This process removes a bit of disorder from the Universe, decreasing the entropy, seeming to violate that fundamental second law of thermodynamics. It was in 1972 that Mexican-born Israeli-American Jacob Bekenstein (then a PhD student at Princeton

University) solved this issue.[114] He realised that as the black hole accretes more matter, and grows in mass, its event horizon also grows. The event horizon is a sphere around the singularity, and so technically that sphere has a 'surface', with a surface area. As the black hole grows, the surface area of the event horizon sphere also grows. It's this area that Bekenstein argued characterises the entropy of the black hole; as it grows the entropy increases and cancels out the lost entropy of the matter falling in. The overall entropy of the Universe still increases, as the second law of thermodynamics decrees.

Hawking, however, wasn't so sure. Entropy is intrinsically linked to the amount of heat energy a process gives off, hence 'thermo'-dynamics. A change in entropy is linked to heat transfer from hot to cold; for spontaneous transfer of heat energy from cold to hot, entropy would have to decrease – it is the least likely thing to occur. This is why a hot drink cools down and cold drinks warm up; heat is transferred from hot to cold as it's the most likely thing to occur. Hawking reasoned that if the surface of the event horizon had entropy then it should be emitting radiation.

Hawking set about to disprove this and he knew to do it

---

114 Bekenstein also developed the 'no-hair theorem' of black holes; that no matter what the black hole is made of on the inside (i.e. what it has accreted over the years), it can be described by three things: its mass, its electric charge and how fast it is spinning. No other information is needed ('hair' being a metaphor for this extra information) to completely characterise the black hole: 'the black hole has no hair'. I guess another way of looking at it would be that black holes don't rely on any hairography to wow us. They are bald.

he would need to tie in quantum mechanics with general relativity. Quantum mechanics is what underpins the behaviour of particles on the smallest scales and it's what gives rise to the laws of thermodynamics. Since general relativity can't help us understand much beyond the idea of a singularity and an event horizon, could a quantum gravity theory help explain what was going on?

In 1973, Hawking visited Moscow to work with Soviet astrophysicists Yakov Zel'dovich and Alexei Starobinsky, who had been applying the ideas of quantum mechanics in the case of extremely curved space, such as around a black hole. They knew that curved space would wreak havoc with the balance of energy in space itself on tiny quantum scales. Much to Hawking's disbelief, they had the mathematics to back up the claim that rotating black holes should be able to create and emit particles, which supported Bekenstein's ideas about a black hole's entropy.

To his annoyance and surprise, Hawking's own initial calculations then showed the same thing (and that non-rotating black holes should also be able to create particles), and it became an obsession to explain what was going on. To explain this fully, you need a theory of quantum gravity; a marriage of quantum mechanics and general relativity to figure out what happens to quantum energy fluctuations in curved space. Unfortunately for Hawking that didn't exist, and it still doesn't in 2022. So instead, he took a shortcut. He considered the quantum energy before and after a black hole had formed when space was and wasn't curved.

The quantum mechanics world is a weird one. There is

energy in space itself, thanks to tiny vibrations, or oscillations to give them their proper physics term. There are certain modes that those vibrations can have; imagine space is a string on a violin and the quantum modes are different notes.[115] Put a finger down on a fret and you'll change the note the string makes (i.e. the energy it vibrates with). Quantum oscillations are a bit different to music notes on strings though, because you can have positive and negative wavelengths which cancel each other out, making a perfect balance of energy (something we call the 'vacuum state').

Hawking argued that forming a black hole in the path of these quantum oscillations could cause modes with wavelengths similar to the event horizon to get disrupted, at which point they would be lost to the black hole. However, other modes of different wavelengths would avoid disruption and continue on their merry quantum way. This would disrupt the balance of energies in the quantum modes in space itself, meaning that some would no longer have another mode to cancel them out. This imbalance in energy gets released as real radiation; light with a wavelength similar to the size of the event horizon of the black hole. So the event horizons of supermassive black holes should emit radiation with a longer wavelength, like radio waves, and smaller black holes should emit radiation with a shorter wavelength, like X-rays or gamma rays, with power that is almost explosive. In fact, Hawking titled his paper describing this process as 'Black hole

---

115 I'm not talking about string theory here, just using strings on a violin as an analogy.

explosions?', although this radiation would eventually come to be known as Hawking radiation.

What was truly remarkable was that when Hawking worked through all the quantum mechanics mathematics to arrive at this conclusion, he realised that the distribution of different wavelengths of radiation given off would be the exact same shape as given off by thermal radiation from something hot like a star. Here again is a link between thermodynamics and the physics of black holes. In everyday thermodynamics, 'black body radiation' is radiation emitted by anything that is heating its surroundings, anything from a star, to an oven, to a human body. Whereas a massive star gives off the majority of its radiation at ultraviolet and optical wavelengths, a human body gives off the majority of its radiation in the infrared, at longer wavelengths. This is because a human is, unsurprisingly, much cooler than a star. For thermal radiation there is a very specific shape to the distribution of wavelengths of radiation given off that is related solely to an object's temperature, a phenomenon that was discovered in 1900 by German physicist Max Planck, one of the pioneers of quantum mechanics. It's why hotter stars are blue and cooler stars are red.

Hawking realised that the radiation produced as black holes disrupted these quantum energy oscillations could be described in the same way, except instead of temperature determining the shape of the distribution, it was the surface area of the event horizon (and therefore the mass of the black hole). Just as Bekenstein had theorised, but had not been able to explain. The big impact of Hawking radiation, though, is that to turn a tiny quantum oscillation into real emitted radiation, some of the energy in that process is borrowed from the black hole itself. Remember, in Einstein's most famous equation, $E = mc^2$, energy and mass are equivalent. So as the black hole loses energy to produce Hawking radiation it also loses mass; the black hole slowly 'evaporates'.

Emphasis there on the *slowly*. Hawking worked out how long this process would actually take, finding that it once again all depended on the mass of the black hole. A black hole the same mass as the Sun could hypothetically eventually evaporate away all its energy as Hawking radiation in a time of 10^{64} years (that's a 1 with 64 zeros after it, or 10,000 trillion trillion trillion trillion trillion years). Bear in mind the Universe itself has only been around for 13.8 billion years and you'll realise just how sloth-like Hawking radiation truly is. Although Hawking did calculate that any primordial black holes that formed in the early Universe with a mass less than 1 trillion kg (for context, the Earth is around 6 trillion trillion kg, so Planet 9 is still safe, don't worry) would have had enough time to evaporate by now.

What's exciting is that if such black holes exist then we might just be able to spot their last gasps of Hawking radiation before they fully evaporate. In the last 0.1 seconds of this evaporation process, a 1 trillion kg black hole would emit the equivalent energy of a 1 million megaton hydrogen bomb. Sounds large, but it's piddly, astronomically speaking; supernovae go off with energies 1 billion trillion times larger than that and radiate for days afterwards.

So while the hope has always been that we'll observe Hawking radiation from a black hole in action, nothing has been detected yet. Hawking radiation remains hypothetical, a great idea on paper but not quite backed up with real data yet. That could just be because we haven't waited around long enough to spot any; the process is so slow that a single human lifetime may not be enough time for radiation that we'd have a hope of detecting to be emitted.

The supermassive black hole at the centre of the Milky Way would be the most likely candidate, but at 4 million times the mass of the Sun, the Hawking radiation will have a long wavelength and be emitted at a much slower rate. It would take 10^{87} years for it to fully evaporate, and that's if it's finished growing and doesn't accrete more material in the future. For TON 618, reaching its maximum mass it can achieve through accretion, the evaporation time is almost 10^{100} years (a googol). Whether the Milky Way's own black hole or TON 618 will ever evaporate depends on how long the Universe will be around for. Does the Universe even have that many years left?

Here at the end of all things

As we draw to the end of this book, it's natural to think about how our Universe might also end. When we look out into space, on average, the light from nearly all galaxies is redshifted. They are speeding away from each other because the Universe is expanding. This discovery in the 1920s led to one of the most famous theories in the whole of science – that of the Big Bang. If you were able to rewind time on the Universe you'd see all the galaxies get closer together and all the matter get squashed down into an infinitely small space. Sound familiar? If you try and put a large amount of anything (mass, temperature, pressure) into an infinitely small space you'll end up with a singularity.

One of the biggest misconceptions about the Big Bang theory is that it is a theory of the creation of the Universe, but it's not. The Big Bang theory describes how the Universe went from an incredibly hot and dense state to evolve to give us the distribution and different shapes of galaxies we see today. It doesn't explain what happens at that very first moment of 'creation' when time = 0. Our knowledge of physics allows us to rewind all the way back to when the Universe was just a scant 10^{-36} seconds old (a trillionth of a trillionth of a trillionth of a second), but before that all our known laws of physics break. The four fundamental forces of gravity, electromagnetism, the strong force (that holds atoms together) and the weak force (that governs radioactivity), behave completely differently and merge into one. To describe those moments we'd need a Grand Unified Theory (a GUT), which we don't yet have. Similar to how Hawking needed a unified theory of quantum mechanics and general relativity to understand the entropy of a black hole, but one didn't exist yet.

So the singularity at the beginning of the Universe is not well understood, but we know it has to be different from the singularity of black holes trapping everything beyond an event horizon, otherwise we wouldn't all be here. For some reason, space started expanding, accelerated by something we call 'dark energy' but we have no idea what that actually is. The physics story is far from complete; there are far more mysteries for budding physicists to crack, standing on the shoulders of all those that have come before them that we've heard about throughout this book.

Just like in stars, the past 13.8 billion years of the Universe's history have been a fight between the expansion of space outwards and the matter in the Universe causing gravity pulling inwards; it's a fight that so far the expansion has won. But if we consider the eventual fate of the Universe, many billions of years in the future, it all depends on how much of the Universe's energy budget went into powering the expansion and how much into making matter. If these two balance each other out then the Universe's expansion will eventually slow, getting infinitesimally small. We have a hope of measuring this with something called the density parameter: the sum of the average density of all matter, radiation and dark energy in the Universe divided by the critical density that would perfectly balance out the expansion. If the density parameter is 1, then the expansion is perfectly balanced by the contents of the Universe and we know that eventually the Universe will reach equilibrium: a happy medium.

If the density parameter is less than one, then matter is out-gunned by the expansion, and the Universe will end in a 'Big Rip' scenario. The expansion will increase exponentially until it doesn't just overpower gravity, but also the strong force binding together the particles in atoms themselves. The Universe would end up as a very sparse collection of lifeless particles.

If the density parameter is more than one, then matter outweighs the expansion. The expansion of space will start to slow before it is reeled back in and starts to contract in a 'Big Crunch'. In this scenario all matter and energy in the Universe would be reeled back together, with pockets of the Universe becoming dense enough to form ultramassive black holes before they too are reeled down into one lone singularity. There's something quite nice about this idea of the Universe being cyclical in nature, bringing the Universe right back to where it started. There are even some astrophysicists investigating the possibility of a 'Big Bounce', where the Universe cycles between Big Bang expansion and Big Crunch contraction endlessly.

To find out which of these scenarios is the eventual fate of the Universe, we can try to measure the density parameter. One of the most accurate measurements we have is from the WMAP[116] satellite observing the radiation from the cosmic

---

116 WMAP stands for Wilkinson Microwave Anisotropy Probe. It's named in honour of American astrophysicist David Wilkinson, who

microwave background, an echo of radiation from the early Universe that reveals what the conditions were like back then. Combining the WMAP data with measurements of the expansion rate of the Universe using supernovae in the nearby Universe, gives a value for the density parameter of 1.02 ± 0.02. That ±0.02 is the uncertainty in the measurement, and means the value could be anywhere in the region of 1.00–1.04.

WMAP revealed that the Universe is tantalisingly close to being balanced, and yet that value errs on the side of matter winning out one day over the expansion. If the value truly is just that teensy weensy bit bigger than 1, then the ultimate fate of the Universe is a Big Crunch. All the stuff in the Universe reeled back into one final singularity: the black hole to end all black holes.

So even as you sit and read this, hurtling through space, happily shepherded around the supermassive black hole at the centre of the Milky Way with no danger of 'falling in', I'm sure, like me, you still can't help but consider the inevitability of black holes. We are intrinsically tied to them in life, and in death our atoms may one day, in an unfathomably distant future, become part of the black hole at the end of the Universe. Let's hope there's a restaurant there too.

---

pioneered the study of the cosmic microwave background through the 1970s. He was a member of the science team for the WMAP project, and managed to see the satellite launch in 2001, but not the new science results it revealed after he unfortunately passed away in 2002 after a seventeen-year battle with cancer.

Acknowledgements

Wow, this book is a LONG one. It feels almost as if it too should be classed as ultramassive. It's a good 61,000 words. For context, my PhD thesis was 56,000 words. I've essentially written another thesis entirely. For someone who was told at school they didn't write very well, I'm still rather in awe of myself. To me, space is hard, but words are harder.

Of course, I've had an entire team of wonderful people behind me to help make this happen. Firstly, thanks to my very first agent Laura McNeill, who pulled this idea from somewhere in the darkest corner of my brain where it was hiding and believed that I could do it. Laura, good luck with all your future endeavours outside of the publishing world. To Adam Strange at Gleam, who then picked up where Laura left off, thanks for being my biggest supporter (although we both know your daughter will fight you for that title).

To all at Pan Macmillan who turned my word vomit of science into a real and tangible book, a huge thank you. To my editor Matthew Cole, thanks for believing in this book from the outset, and for pointing out all the bits of the puzzle that were still missing for the black hole uninitiated. Thank you to Charlotte Wright and Fraser Crichton for going through this manuscript with a fine-tooth comb to correct all of my neglected grammar and sentence structure. And of course to Josie Turner, Jamie Forrest and the whole team at Pan Macmillan for all their efforts publicising and marketing my book to the big wide world.

My sister, Megan Smethurst, is the wonderful human being responsible for the diagrams throughout this book. While I'm the scientist, she is the artist in the family and I will forever be completely in awe of her talent. I now have many more opinions on fonts than I did before – thanks Meglar.

A huge, reverent thank you to all those scientists who came before me for all their efforts, and especially to the women before me who blazed a trail through the scientific world of men so that my role as an astrophysicist is not questioned today.

This book was written in the back half of 2021, when the world was just starting to go back to normal, and finished just before Christmas 2021 when that normality was once again in danger. It has been written across cafes, offices and homes; in particular one 'writing retreat' staycation week in Cambridge. It was there that I realised that the Cavendish Laboratory holds *that much* of the history of black holes, which led to the exasperation of footnote 49. Thank you to all cafe owners and workers who recreated that long-missed buzzy environment for academics like me wanting a new space to work in and provide inspiration.

Along with the professional army, there's also a personal

army supporting the writing of any book. To Mum, Dad and, once again, my sister Megan, thanks for always believing in me and for the excitement you all expressed to finally read it. I love you guys so much and know that collectively you will get all of the random pop-culture and lyric references that I named my chapters after. 'Standing on the shoulders of giants' is an Oasis quote, right?!

Speaking of lyrics, I also motivated myself with music on many evenings I found myself writing after my regular work day of research was done. There are at least three Taylor Swift references through this book; her music and lyrics resonate with me so much and I will forever be in awe of those who can create something so powerfully beautiful, like she does. *Folklore*, *Evermore* and *Red (Taylor's Version)* were the main soundtrack to my frantic typing.

Finally, the words thank you don't do my partner Sam justice. 'Frodo, Becky wouldn't have got far without Sam.' For every long evening spent writing, for listening while I regaled you with all the 'fun facts' I learnt while researching, for every smile that picked me up at the end of the day, thank you. I love you, *always*.

Bibliography

Emilio, M., et al., *ApJ*, vol. 750, p. 135 (2012)

Giacintucci, Simona, et al., *ApJ*, iss. 891, p. 1 (2020)

Huygens, Christiaan, *Treatise on Light*, translated by Silvanus P. Thompson, www.gutenberg.org/ebooks/14725 (1678)

Kafka, P., *MitAG*, vol. 27, p. 134 (1969)

Manhès, Gérard, et al., *Earth and Planetary Science Letters*, vol. 47, iss. 3, p. 370 (1980)

Montesinos Armijo, M. A. and de Freitas Pacheco, J. A., *A&A*, vol. 526, A146, doi:10.1051/0004-6361/201015026 (2011)

Rindler, W., *MNRAS*, vol. 116, iss. 6, p. 662 (1956)

Röntgen, W. C., 'Ueber eine Neue Art von Stahlen', *Sitzungsberichte Der Physik..-Med Gesellschaft Zu Würzburg* (1896)

Scholtz, Jakub and Unwin, James, *Physical Review Letters*, vol. 125, iss. 5, 051103 (2020)

Schwarzschild, 'Letter to Einstein', *Schwarzschild Gesammelte Werke (Collected Works)*, ed. H. H. Voigt, Springer, 1992, vol. 1–3 (1915)

Webster, L., Murdin, P., *Nature*, vol. 235, iss. 5332, pp. 37–38, doi:10.1038/235037a0 (1972)

Wheeler, J. A., *AmSci*, vol. 56, 1 (1968)

Index

Page references in *italics* indicate images.

INDEX